Behavioral Computational Social Science

WILEY SERIES IN COMPUTATIONAL AND QUANTITATIVE SOCIAL SCIENCE

Embracing a spectrum from theoretical foundations to real-world applications, the Wiley Series in Computational and Quantitative Social Science (CQSS) publishes titles ranging from high-level student texts, explanation, and dissemination of technology and good practice, through to interesting and important research that is immediately relevant to social/scientific development or practice.

Other Titles in the Series

Behavioral Computational Social Science

Riccardo Boero

Los Alamos National Laboratory,
New Mexico, USA

WILEY

This edition first published 2015
© 2015 John Wiley & Sons, Ltd

Registered Office
John Wiley & Sons, Ltd, The Atrium, Southern Gate, Chichester, West Sussex, PO19 8SQ, United Kingdom

For details of our global editorial offices, for customer services and for information about how to apply for permission to reuse the copyright material in this book please see our website at www.wiley.com.

Library of Congress Cataloging-in-Publication data applied for

A catalogue record for this book is available from the British Library.

ISBN: 9781118657300

Cover Image: maikid/iStockphoto

Set in 10.5/13pt Times by SPi Global, Pondicherry, India

Printed and bound in Singapore by Markono Print Media Pte Ltd

1   2015

Contents

Preface

I developed the ideas and perspectives presented in this book during several years of work and interactions in many different places and with many people. I should probably thank all of them personally and individually and not only for what I present here.

Among the many institutions I should thank, there are the Universities of Torino and Brescia in Italy, the University of Surrey in the United Kingdom, and the Los Alamos National Laboratory in the United States. I have had the chance to "live" in those institutions, and I consider the seminars, the talks, and the discussions I had in those places with both staff and students as probably the most inspiring ones for the development of my research interests and approach.

I also have to admit that the fact of having worked for several years in the field of public policy has helped me greatly. When scientists face real-world problems, they can learn much, both because they have the chance to put their knowledge to use in a real-life situation (and to fear the results) and because they receive unique feedback from policymakers and stakeholders. I want to thank in particular those I met in European local governments who shared with me their own viewpoints, unconsciously supporting my learning.

Among the many individuals to whom I owe a thanks, there are, in a disordered order, Pietro Terna, Flaminio Squazzoni, Nigel Gilbert, Peter Hedström, Paul Ormerod, Laura Bianchini, Brian K. Edwards, Marco Castellani, Giangiacomo Bravo, Gianluigi Ferraris, Matteo Morini, Michele Sonnessa, Donatella Pasqualini, Marco Novarese, Salvatore Rizello, Massimo Egidi, Marco Monti, Rosaria Conte, and finally (and collectively) the students of the PhD program in economics and complexity at the University of Torino. I consider several of them not only fellow researcher but also friends, with some of them I have spoken for entire years, while with others I have had the chance to interact only for short periods and few words. However, each one of them has taught me something special about science in general and social sciences in particular, and for that, I am very grateful.

The points of view presented here are obviously my own only. Because I conceive social research as a never-ending process of innovation and learning of new methods, I hope that this book will contribute to new explorations and ideas.

1

Introduction: Toward behavioral computational social science

The main assumption of this book is that individual behavior and social phenomena are somehow connected and that the investigation of that connection is central for all social sciences.

The work presented here can be classified as a methodological one since it deals with methods. With extreme synthesis, it presents the methods available for putting together the studying of individual behavior, as developed in behavioral sciences, with the many tools that today compose the approach called "computational social science" (from now on CSS).

The ideas and methods presented here have originated in different domains, and it is very difficult today to find an exhaustive and comprehensive description of them. The book thus aims not only at theoretically discussing a unified methodological approach but also at providing the readership with all the necessary information to experiment with the approach. Obviously, given its physical constraints, the preparation of this book has meant much selection and not all concepts and tools are explained from scratch and in details.

Behavioral Computational Social Science, First Edition. Riccardo Boero.
© 2015 John Wiley & Sons, Ltd. Published 2015 by John Wiley & Sons, Ltd.

However, average readers with interests in the scientific explanation of social phenomena can surely comprehend what is discussed here and can use the many references provided to master all the topics and tools.

Before presenting how the book is organized and how selective reading can be conducted on it, this introduction focuses on the first question readers should ask: what is the use of this approach?

1.1   Research strategies in CSS

CSS has recently emerged because of many important technological and conceptual advancements.

Undoubtedly, it is the "Big Data" approach and the large availability of data associated with it that allow today studying large-scale social phenomena that were impossible even just a few years ago. At the same time, it is the availability of (cheap) computing power that allows storing, managing, and analyzing those datasets.

From the conceptual viewpoint, relatively new scientific tools such as those of social networks analysis, complexity, and other approaches finally find in such an abundance of social and behavioral data the chance to be applied and tested.

Along with the promises of the potentially fruitful integration of all those innovations, CSS seems today to give social sciences the possibility to overcome the well-known limits of more traditional approaches. Heterogeneity of individuals, nonlinearity of systems and behavior, and the lack of capability to effectively put in relationship social structures and social phenomena with individual behavior are just a few of the many examples of limits that potentially can be overcome.

Much of the research in CSS is today still aimed at exploring the potentials of the approach, but different research strategies have already emerged, and consequently, different methodologies have emerged too.

In particular, CSS can be considered today as a self-standing approach to social sciences because it provides tools and methods to pursue any kind of scientific research strategy.

Research strategies in science are only a few. Following an order that is not intended to imply any ranking of importance, it is firstly possible to explore the data in order to describe or classify it. It is an activity that is always needed in scientific investigations and that can provide first-hand and novel information about completely unknown phenomena and systems.

Second, it is possible to establish relationships in the data. Using statistical and other models, it is in fact possible to observe that some of the variables appear to be connected, changing in similar or opposite ways. Variables that are somehow related are the first candidates to consider, select, and investigate deeper.

Most of the contemporary research works in CSS, for instance, the ones belonging to the Big Data and science of networks approaches, adopt one or both of these two research strategies.

CSS however allows also pursuing the third kind of scientific research strategy, which is the investigation of causality. Similarly to any other scientific domain, in social sciences, the investigation of causality requires the availability not only of data but also of tools for modeling. Modeling is the formalization procedure that ultimately allows developing, testing, and validating knowledge. In CSS, modeling is pursued by the approach of social simulation where adopted modeling tools are explicitly addressed at dealing with the peculiar complexities of social systems.

1.2   Why behavioral CSS

Behavioral CSS is aimed at investigating causality of social phenomena and it thus relies on modeling tools developed in social simulation.

Further, it refers to the need for behavioral information. The need for the integration of the two approaches is motivated by the unique form that causal relations have in social systems.

In fact, social phenomena can be conceived as different from phenomena in natural systems because of their complexity. The complexity of social phenomena, moreover, is surely characterized by the most complex object we know, which is the human brain, but not only by that. Social phenomena, in fact, are complex because their causes always involve both individual behavior and some of the many features of the social structure (e.g., institutions, social norms, ways of social interaction, etc.). These features of social complexity are the ones that the behavioral CSS approach explicitly acknowledges and that it tries to effectively deal with.

Most of the critiques to established research methods in social sciences are addressed at the same time. Causal explanations in traditional approaches are incomplete and often unreliable because the complexity of social systems is reduced, either from the perspective of individuals or from the one of the social structure. Established approaches are often obliged to apply reductionism because of the technical limits of the analytical tools that are adopted.

Today, CSS and behavioral sciences create the opportunity to overcome traditional limits and to finally deal with the individual, the social structure, and the relation between the two.

Behavioral CSS can thus be seen from three different perspectives. From the one of behavioral sciences, it can provide researchers with the tools to extend their investigations toward social contexts of interaction. Such an extension can improve knowledge about behavior because of the many feedbacks going from the social structure to individual behavior and *vice versa*.

From the perspective of researchers in social simulation, the integration of behavioral knowledge in their models can finally allow rigorous validation and external validity.

From the perspective of social sciences in general, behavioral CSS enriches the researcher's toolset with an approach that is explicitly addressed at the investigation of causality of social phenomena intended as a social causal mechanisms.

Because of what just said, the behavioral CSS approach presented in this book is aimed at readers in social sciences in general and particularly at those in behavioral sciences and social simulation.

The book discusses the approach and presents several tools in order to allow readers with different backgrounds experimenting with and perhaps extending it.

1.3   Organization of the book

Being aimed at presenting and discussing a methodological approach that puts together different tools and traditions, the book is organized in two parts. In the first one, there is the presentation of the main concepts and methods developed in the two approaches that are integrated (i.e., CSS and behavioral sciences). The first part also includes a short discussion of the advantages of the approach.

The second part takes a more applied and technical perspective and it presents methods for tool integration. In particular, because in order to investigate causality with behavioral CSS researchers have to rely on modeling, that part initially discusses how to integrate results of behavioral analyses in models. Secondly, there is a chapter that discusses how to model structures of social interaction.

In conclusion, the second part ends with the presentation of an example of application of the approach, mainly aimed at didactic purposes. In particular, it is presented the preparation and implementation of a model of a social phenomenon, starting from the collection of behavioral data, passing through its analysis, and arriving at the specification and formalization of behavior and interaction in the model. Some applications of the model are sketched too in order to give the reader the intuition of potential uses and results in terms of causal explanations.

Readers experienced in social simulation and agent-based modeling can probably read the book without spending much attention to Chapter 2, although the concepts related to causality that are presented there are crucial for the evaluation of the approach.

Readers coming from behavioral sciences can avoid Chapter 3 where common tools of that approach are shortly introduced.

Readers particularly doubtful about the analytical advantages provided by the approach should start reading Chapter 4 and then read the section presenting results in Chapter 8. If reading those chapters changes their minds, they can go backward to the rest of the book to better comprehend the reasons why the approach can be analytically effective and powerful.

Readers who intend to adopt the modeling approach presented here but who are not experienced in social simulation should spend particular attention to the second part of the book. The first three chapters of that part illustrate the algorithms that allow modeling behavior in agents and interaction between agents. The fourth chapter of that part provides examples of application of many of those algorithms. Finally, in the appendix, the technical implementation of what presented in Chapter 8 is reported and discussed.

Part I

CONCEPTS AND METHODS

2

Explanation in computational social science

What is computational social science (CSS) and how researchers working in this field think and work? This chapter aims at shortly introducing the main concepts, methods, and tools underlying the emerging field of CSS.

CSS is an intrinsically interdisciplinary approach that uses several concepts originally developed in different domains. Furthermore, its development takes also advantage of the availability of innovative modeling tools, of the increasing computing power of computers, and of everyday larger and detailed datasets about social phenomena.

In order to understand CSS, it is thus needed to understand the convergence of concepts and methods that constitutes it. Further, while the CSS approach is today made by several different tools applied to different research strategies, for the sake of this work, we focus only on one specific research strategy, that is, the investigation of causality in order to explain social phenomena. Therefore, we select concepts and tools that have influenced the development

Behavioral Computational Social Science, First Edition. Riccardo Boero.
© 2015 John Wiley & Sons, Ltd. Published 2015 by John Wiley & Sons, Ltd.

of CSS with that in mind, and we focus our discussion on how scientific explanations (i.e., investigations of causality) are conceived in CSS.

The first section of this chapter is devoted to the introduction of the main concepts underlying CSS, namely, the ones derived from the fields of complexity, social mechanisms, and social networks. They all contribute to the definition of CSS because they provide CSS with concepts, methodologies, methods, and tools. Furthermore, the section outlines the contributions of two other important elements influencing the development and definition of CSS, which are the increasing availability of data regarding socioeconomic domains and the increasing computational power available to researchers.

The following section then focuses on the methods used in CSS and particularly on the predominant role played by agent-based modeling, network analysis, and social simulation in general.

The fourth section of this chapter is a very short introduction to the general and common characteristics of the software tools available for conducting causality research in CSS, aiming at allowing inexperienced readers to evaluate research needs and tools potentials before tackling steep learning curves.

The chapter concludes with a short discussion of some important practical critical issues related to the CSS approach to explain social phenomena.

2.1   Concepts

CSS derives from the convergence of concepts and tools, and concepts and tools themselves in turn are often the output of other innovations such as the availability of new large datasets and computing power.

The understanding of concepts underlying CSS is thus necessary in order to understand what CSS is and to more easily appreciate contributions in this field. However, it is hard to defend that each single constituent of CSS is a novelty. On the contrary, disciplines composing social sciences have debated and developed the same concepts underlying CSS for decades if not centuries, and they have done that often in autonomous and independent way, in different times, under different labels, and with different impacts. The novelty and the definition of CSS thus do not descend from a single concept, but from the adoption of a broader, consistent, and inclusive perspective.

2.1.1   Causality

The first important contribution to the development of CSS derives from complexity science, that is to say the science that studies complex systems (Eve et al. 1997, Foster 2004).

Complex systems are systems defined as, in general, characterized by the presence of nonlinearity and by being far from equilibrium. Nevertheless, they are not chaotic systems, and they have several properties, like the possibility to generate outcomes difficult to forecast or understand by just looking at the elements of which they are composed.

In complex systems terminology, there is a micro level which is made of system components and a macro level that is made of aggregated and systemic outcomes. A good and classic example is the temperature of materials, a macro feature absent in the micro level of atoms.

The research field coping with complex systems was developed within physical sciences, in particular in the fields of nonlinear and nonequilibrium thermodynamics (Prigogine and Stengers 1979). More recently, it has been extended to social sciences.

The modeling frameworks developed in complexity, for instance, in the field of statistical mechanics, mostly rely on simulation techniques. Nonlinearity in fact often means the absence of closed form analytic solutions, and thus, simulation is the unique way to explore the solution space and to understand model outcomes.

Complexity thus carries on the attention toward nonlinearity, on out-of-equilibrium dynamics rather than on static equilibria, in the usage of simulation as the analytical tool to investigate models of systems that can be out of equilibrium, and finally on the presence of macro, systemic outcomes that "emerge" from micro heterogeneous components. In other words, the approach provides concepts and tools to investigate causality in complex systems.

Regarding the focus of complexity on emergent systemic properties, one of the most known and studied of them is self-organization. Complex adaptive systems (Holland 1975, 2006) are a subcategory of complex systems capable to organize themselves autonomously to cope with external shocks and to adapt to a changing environment. Any kind of living organism is a good example of a complex adaptive system. Most social systems are good examples too.

Another strand of research that has influenced the development of CSS is the one focusing on social causal mechanisms as explanations for social phenomena. This field of research provides CSS with a specific form of explanations (i.e., mechanisms) for social phenomena. Social mechanisms are, in other words, how causal links are organized and have to be searched for in social systems.

Definitions of social mechanisms are many. For instance, Merton, who introduced middle-range theorizing as the middle way between general social theories and descriptions, proposed mechanisms as constituents of middle-range theories. Mechanisms in his words are "social processes having designated consequences for designated parts of the social structure" (Merton 1949, pp. 43–44).

The relation between social processes and the social structure helps pointing out also the presence of different levels in social structures, so that mechanisms work between "entities at different levels, for instance between individuals and groups" (Stinchcombe 1991, p. 367).

Mechanisms, however, can be defined also via their explanatory power. For example, Harré (1970) noted that a mechanism needs to have a key role in explaining a social phenomenon, while Elster (1998) reports the evolution of his interpretation of mechanisms, saying, for instance, that in Elster (1983) he conceived mechanisms as antonyms of black boxes, in connection with a reductionist strategy in social science, while in more recent works, he was referring to mechanisms as antonyms of scientific laws. Mechanisms, in summary, have to be white boxes because they are explanations of the causality of a social phenomenon, but they take a specific form that is far from the one of scientific laws.

Combining the ideas of mechanisms as social processes and as explanations of social phenomena, Boudon (1998, p. 172) presents the definitive definition of a social mechanism as the "well articulated set of causes responsible for a given social phenomenon." The need for mechanisms derives from the characteristics of causality in social phenomena. Other kinds of explanations of social phenomena (e.g., laws) do not allow effectively capturing causality of these systems due to their reductionism.

Further, social mechanisms do not imply a mechanicistic and static vision of social systems, but on the contrary, they suggest one that shows features like self-organization, self-adaptation, and purposive behavior as in complex systems. In fact, as early pointed out for neural cognitive mechanisms in human beings, there are also mechanisms capable to (purposely or not) adapt the operations of other mechanisms to produce different results in the same external conditions (Hayek 1952). Similarly, social mechanisms define social systems where the internal structure can be modified and the range of operation can be extended depending upon experience.

Explanations in the form of social mechanisms are not in the form of a law (either deterministic or probabilistic). Social mechanisms also differ from explanations in the form of statistical relationships because they are made by causal relationships of which statistical correlations are only an observation of some outcomes.

Social mechanisms, that have found a systematic representation in sociology in the approach of analytical sociology (Hedström and Swedberg 1998a), are not new in social sciences, but they support CSS in identifying two main aspects of the scientific explanations looked for in this approach.

The first one is the focus on causality: in CSS, it is essential to investigate the complex set of causes that generate social phenomena. The degree of complexity of causes cannot be reduced and must be effectively coped with. Consequently, CSS mainly exploits modeling tools that allow rigorously investigating the complex set of causes giving rise to social phenomena.

The second consequence of the social mechanism approach for CSS relates the attention on the distinction and relationship between micro and macro.

Social mechanisms consider two sets of abstract elements, actors, and their "staging," and they are useful because they allow understanding their systemic effects (Hernes 1998). The distinction between "macrostructural and microindividual levels" (Van Den Berg 1998) allows investigating both how social structure influences individual behavior and how individual behavior determines social structure.

The distinction between micro and macro is thus not just a trivial separation between individuals and the aggregation. For instance, social mechanisms allow for the possibility of considering institutions and other kinds of actors as micro entities. Similarly, explanations neither have to be reduced to methodological individualism (i.e., to individual actions) nor to social structures. Social mechanisms neither ignore the relevancy of actors nor of structures, but they consider both as necessary ingredients for understanding social phenomena.

The micro–macro separation is thus dependent on the phenomenon of interest and on the elements of analysis, and it is led by the fact that some mechanisms needed to explain the phenomenon work between two conceptual different levels of social systems, one of which is greater than the other in the sense that it contains the other. For instance, if the phenomenon of interest is a particular dynamics happening in a society, the macro level could be that social group, with all its characteristics, rules, and so forth, while the elements aiding the description of the macro level could be the elements of the micro level, that is to say individuals and their relationships.

In summary, the distinction considered in social mechanisms between the micro and the macro is characterized by the following features:

- It is a relative, nonabsolute separation depending on the social phenomenon on focus and on the research process carried on by the scholar studying it.

- It considers a micro level constituted by elements (from individuals to nations) who act and interact.

- It considers a macro level that represents the system made by the micro elements, which is often conceived as a structure (which can be composed of roles, social norms, etc.) emerged from social actions by micro elements: the only requirement is that the features of interest of such a "macro" system must be observable (i.e., some features of the macro must be measurable).

• The micro and the macro are separate for analytical purposes, but for the same reason, they are put in relationship for explaining social phenomena: social mechanisms in fact often point out the effects of the macro on the micro and how the micro generates the macro.

An abstract model can be used to synthesize and categorize how explanations of social phenomena usually work with social mechanisms and in CSS. In particular, it is possible to consider a modified version of the macro–micro–macro model introduced by Coleman (1986), often dubbed as the "Coleman's boat." The aim of the model is to represent three different kinds of causal links determining social phenomena. The different kinds of causes are separated here for the sake of simplicity and for improving the description and the modeling of social phenomena (see Section 2.3 when talking about tools used in CSS).

As in Figure 2.1, which schematically presents the abstract model inspired by Coleman, the framework considers micro and macro levels and their reciprocal influences. As the figure shows, social mechanisms can be differentiated depending on the direction of the connection between macro and micro levels. In the figure, numbers represent causal relationships that contribute to explain a social phenomenon, and the direction of the arrow indicates the direction of causality.

Causal relationships where the macro level influences the micro elements are represented in Figure 2.1 with the arrow labeled "1." They include dynamics such as "situational mechanisms" from analytical sociology (Hedström and Swedberg 1998b), "second-order emergence" from social simulation (Gilbert 2002), and "immergence" from the fields of cognitive modeling and social simulation (Castelfranchi 1998).

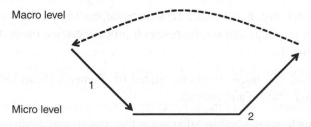

Figure 2.1   An abstract model of causal relationships in computational social science.

For instance, many belief-formation and preference-formation models have at their core the presence of mechanisms which link the social structure or other macrosocial characteristics to the beliefs, desires, opportunities, and behavior of actors.

In the field of economics, a common example is the influence that macro-economic variables such as the official interest rate can have on households and firms beliefs and choices.

Another example of this kind of causal mechanism is constituted by the complex set of constraints and opportunities faced by individuals in any kind of social dynamics. In fact, the environment that surrounds individuals, either the social structure or the natural environment, is a macro-level entity that defines opportunities and constraints of actors living there.

Similarly, another example could be the one of some learning processes where the feedbacks do not come from other micro actors but from the macro level. Moreover, a set of social norms can influence beliefs and actions of individuals, in particular when individuals have low levels of reflexivity or when social identity is strong.

The second arrow in Figure 2.1, the one labeled "2," describes causes entirely located at the micro level and causes that affect the macro level by means of interaction and aggregation at the micro level.

They include dynamics happening at the micro level that do not involve interdependence with the macro level and also those that do not strictly require social interaction, as, for instance, utilitarian decision-making. But they also include any other kind of action and choice at the micro level if not influenced by the macro one.

Thus, many of those causes are based upon interaction among micro elements, including processes such as imitation and communication. Social interaction, moreover, often defines how micro elements come to define the macro level or to modify it.

Along the arrow labeled "2" takes place all the processes of social interaction that in CSS often relies on the concepts developed in social network analysis (SNA).

Starting from some seminal works of social psychologists and sociologists (e.g., Milgram 1977), SNA is a branch of sociometrics that has developed tools explicitly aimed at observing, measuring, and statistically investigating the structure of individuals' interaction. In recent times, SNA has experienced a renewed attention due to the availability of vast datasets about social interactions on social media and by scholars in the field of complexity.

CSS researchers use SNA to empirically study and describe social networks with the aim of understanding and realistically modeling social interactions.

SNA includes a vast number of statistical measures (see Wasserman and Faust 1994) that can inform the modeling of social interactions and thus that are crucial to investigate social mechanisms (see Chapter 7).

The scheme presented in Figure 2.1 presents also a dashed line that points out the fact that because the macro level is defined by what happens at the micro one, we can have loops where the macro influences the micro (arrow 1) and the micro modifies the macro (arrow 2).

However, what is not represented in the figure is the timing of processes. Take, for instance, the example of a social phenomenon that happens because of daily interactions among individuals and that is also influenced by social norms. Arrow 1 includes the influence of social norms on individuals' choices; arrow 2 includes individuals' interactions and choices and how these determine the social phenomenon on focus. However, social norms are also an output of individuals' interactions and choices, but they do not evolve so quickly that their dynamics enters the explanation.

In summary, an essential point to interpret Figure 2.1 but also to evaluate any kind of social mechanism is the issue of time and of the coexistence of dynamics with different timings and speeds.

A second critical point of Figure 2.1 is that dynamics separated in the two solid arrows are often mixed in reality. For instance, an individual belief can be influenced both by the macro level and by the interaction with another individual. The point is that the figure does not imply that CSS scholars reduce the intrinsic complexity of social phenomena by separating processes according to the direction of the interdependence between analytical levels, but only that they always take into consideration the interdependence among them.

Explanations of social phenomena as in social mechanisms are not completely new to social sciences. The uniqueness and novelty of CSS is to put these concepts together, to use a methodology consistent with them and, probably for the first time in history, to have at disposal enough data to empirically ground social explanations in this form.

There are several important examples and applications in social sciences of the concepts introduced above that predate the emergence of CSS. Starting from economics and from the attention toward out-of-equilibrium dynamics and complex systems, many scholars in this field have extensively refined the concept of equilibrium to take into account a more dynamic perspective and to analyze the space of solutions beyond equilibria.

Moreover, the discipline has always been pervaded by methodological individualism, and even in classical authors, it is possible to recognize several ideas in line with the complexity approach. For instance, Mill (1909 [1848]) and Hayek (1945) have posed heterogeneous information dispersed in economic agents at the basis of their interpretation of efficient markets (and of spontaneous order), and their idea of dispersed market economies resembles the concept of complex adaptive systems from many viewpoints.

Similarly, even in branches of the discipline where methodological individualism and the attention toward the micro–macro link have been historically weaker, such as macroeconomics, today, that is not true any-more (consider, for instance, the effort in microfounding macroeconomic models after Lucas' critique).

If compared with CSS, the traditional approach in economics presents too many constraints in the realism of behavior, heterogeneity of agents, and social interaction. Such constraints, which are most common targets of het-erodox approaches (Davis 2004, Hargreaves 2004), are often criticized of being motivated only by the set of tools adopted and by their characteristics. CSS, as detailed in the following section, addresses these concerns with a set of methods providing the capability to investigate causality in the form of social mechanisms.

Even in classic sociology, many mechanism-based explanations of social phenomena can be found. For instance, in the Weberian Protestant ethic, a cultural issue, at one moment in time, affects people's behavior, giving rise to unintended strong changes in economic activities. Other examples of mechanism-based explanations can be found in the works of Simmel, Durkheim, and Elias (e.g., Elias 1991 [1939]).

Many more recent examples of the use of social mechanisms in sociology can be found, and there are also some well-known generic social mechanisms and families of mechanisms (i.e., "mechanisms that produce similar results, and enjoy similarities but also differences"—Schelling 1998, p. 38).

One family of mechanisms dates back to the work of Merton (1949), and it is usually called the "self-fulfilling prophecy." Many examples can be found in several empirical cases (from coffee shortages to insolvent banks) that resemble the general case, that is, that "an initially false definition of a situation evokes behavior that eventually makes the false conception come true" (Hedström and Swedberg 1998b, p. 18).

Similarly, there are "network diffusion" mechanisms as identified in the works of Coleman et al. (1957, 1966). The diffusion of new products, habits, values, and so on is strictly influenced by the social network structure and by the fact that individuals are more open to adopt innovations if other neigh-boring individuals have already done so, making clear the true value of the innovation.

A third and final example of families of mechanisms is the Granovetter's "threshold theory of collective behavior" (Granovetter 1978). Threshold-based behavior is when the decision of an individual to join a group for a collective action or to behave in a particular manner is dependent on the number of other individuals who have already joined the group or started to behave in that way. Individual heterogeneity of thresholds, as well as the different structure of the problem, gives rise to many different threshold-based mechanisms that can explain many empirical phenomena, from choice of restaurants to dynamics in minority games.

Finally, it is worth adding that in social sciences in general, the development of critical realism (e.g., Fleetwood 1999) has lead to the acknowledgment of the importance of linking micro and macro levels in a causal way and to an early adoption of the concept of emergence derived from philosophy rather than from complexity.

2.1.2   Data

Another essential ingredient of CSS is data. CSS, in fact, is a data-driven social science, and the reason for that is twofold. Firstly, there have been significant improvements in the collection of data about social phenomena, and, for instance, today, it is available data about social interaction through social media. Secondly, social data now is often "open."

Developments in social data collection have been made possible by technological innovations. Data collection methods in fact have changed. For instance, today, much of official surveys administered by census offices around the world are based on methods such as continuous surveys (e.g., the American Community Survey), and thus, the frequency of data collection has changed drastically, often becoming annual and quarterly. This shift is radical, if compared with previous practices that administered surveys every 10 years or 5 at best.

Moreover, researchers and companies interested in collecting original primary data have extended the advantages of computer-assisted interviewing to the web and social media.

Talking about social media, they are a groundbreaking innovation for CSS, *per se*. In fact, being based on computer-mediated communication, they allow what has never been possible before, which is to have precise and extensive measurements of social interaction among individuals. The data they provide is one of the reasons of the renewed interest toward social network statistics in CSS, though it also requires a renovated attention toward ethical consequences of its use.

Lastly, data today is often open. The open data policy adopted by several governments and governmental agencies has further increased the availability of important information.

2.2 Methods

By adopting explanations in the form of social mechanisms, CSS scholars investigating causality are extensively engaged in modeling. Modeling requires extensive computing power considering the use of large amount of data for calibration and validation of models. Models are made by complex structures and dynamics and require numerical simulation and sensitivity analysis. The technological evolution of computers has thus played a fundamental role in CSS, and it is the reason for the "C" of the acronym, that is, for being a "computational" social science.

The main modeling tool used in CSS is agent-based modeling. It is the preferred one because of its flexibility that allows capturing all the epistemological concepts and methodological requirements introduced in the preceding section.

Sometimes, other modeling tools are used by CSS scholars: they will be discussed after agent-based models (ABMs), pointing out the reasons for preferring agent-based modeling when acknowledging the importance of social mechanisms as explanations of social phenomena. Since now, and most importantly, it is worth underlining that all these tools, agent-based modeling included, rely on simulation techniques to be solved and analyzed.

2.2.1 ABMs

ABMs are formal tools for scientific modeling, derived from the tradition in microsimulation (for an illustration of their historical development, see Gilbert and Troitzsch 1999, p. 158 ff).

Their development overlaps with similar research agendas, in particular with the one of multiagent systems (MAS) in computer science, which aims at developing autonomous software capable of interacting with other software and with human users.

They are constituted by autonomous software pieces that are called agents. In modeling social systems, they usually represent individuals, but they can also represent institutions, organizations, and so forth.

ABMs allow multilevel analyses, not bounding the number of levels (i.e., it is also possible to model more than two levels). Any kind of interaction structure can be modeled, and it can be static or dynamically generated and modified by agents.

They are quite an open framework: except for the presence of interacting agents (i.e., agents that interact each other and/or with the environment), there are no other particular prescriptions characterizing them.

ABMs are formal models based on two main elements, which constitute, all together, the so-called model "micro specification." They are (i) a set of agents, with their own resources and autonomous behavior, probably heterogeneous, and (ii) an interaction structure, static or dynamic, independent, or determined by agents' behavior.

Defining those two elements through a programming language constitutes the model. Its execution (often referred to as the simulation of the model) allows obtaining the "macro results" of the formalization process (Squazzoni 2012).

It is important to underline that ABMs results are often called the "macro" side of the model, and the specification is called the "micro" one. Generally speaking, the model is not only built by its specification but also by its results because they formally derive from the micro layer. Moreover, the distinction between micro and macro derives from the fact that ABMs are built bottom-up, that is, obtaining an emerging macro result (upper level) by the interaction of micro elements (lower level).

ABMs do not pose limitations on multidisciplinarity, as they do not on heterogeneity and behavioral modeling. Following Epstein (1999), ABMs are not multidisciplinary *per se* but, because of their flexibility, it is easy to get over disciplinary boundaries traditionally imposed.

Further, the use of ABMs does not imply any particular and specific methodological change from standard scientific practices. Talking from the historical perspective of the development of social simulation and of the evolution in the use of ABMs, a long-lasting methodological debate has involved scholars in the field on whether ABMs call for different scientific methods.

Proposers of a brand new approach asserted that ABMs should be somehow used to study reality in a different way from when inductive and deductive tools are used (Axelrod 1997b, Epstein 1999), that their validation should be based on the principle of "generative sufficiency" (i.e., a model is valid if capable of mimicking a realistic macro behavior; Epstein 1999), and that agents' behavior should be modeled according only to simplicity (i.e., according to the army motto KISS—keep it simple, stupid; Axelrod 1997a, Epstein 1999, Shubik 1996). According to this latter methodological standpoint, ABMs should be unrealistic models in their micro specification but capable of generating outcomes similar to real ones (Axelrod 1997b). However, today, there is no methodological argument to support such positions since ABMs are flexible and they can be used according to any approach.

Pioneering and exploring research in social simulation has in fact pointed out ABMs flexibility, and it has not found technical limitations in their usage that ask for specific new methodological approaches. Further, results from both social simulation and behavioral sciences have soundly shown how behavior should not be reduced to simplicity and unrealism. The empirical salience of the micro specification and of agents' behavior is crucial to achieve a causal explanation of social phenomena, and thus, simplicity in behavior, besides not uniquely defined (Popper 1935), is not supported by the Occam's razor. Furthermore, from the validation viewpoint, the capability of generating realistic macro outcomes is a necessity and not a sufficiency.

However, acknowledging the flexibility of ABMs does not deny the value of using ABMs based upon unrealistic behavior for specific research purposes. For instance, the research agenda developed in economics within the approach called "zero intelligence economics" (Alchian 1950) is focused on demonstrating that perfect rationality is not needed to reach equilibria and to observe other features of markets. Research agendas like that one can be facilitated by exploiting ABMs (e.g., Terna 2000).

In summary, ABMs are a formal tool for modeling social systems, and they should be used according to established standards. However, they can be considered "as a 'third way' of carrying out social science, in addition to argumentation and formalization" (Gilbert and Terna 2000) if by formalization we mean other, more common and traditional, tools. In fact, agent-based modeling is obviously different from argumentation because it involves the usage of rigorous formal modeling, and it is also different from the traditional way of formalization in social sciences that is based on systems of equations.

Considering ABMs just as a new tool means having a richer toolset in which the new language available is complementary to the verbal one and the traditional mathematical one (as suggested in Ostrom (1988) and Gilbert and Terna (2000)).

The last methodological point related to ABMs worth being examined in this section is the process of understanding dynamics and causal mechanisms inside a model.

Besides the relevant issues of model verification (i.e., if the code implementing the model does not contain bugs) and validation (i.e., if the model has empirical salience), the literature in the field of artificial life pointed out one ABMs peculiarity related to the complexity of the modeled processes. ABMs in fact have been defined as "opaque mental experiments" (Di Paolo et al. 2000), that is to say "thought experiments in which the consequences follow from the premises, but in a non-obvious manner which must be revealed through systematic enquiry."[1]

---

[1] Mental experimenting is just one of the possible uses of ABMs (Bedau 1998, 1999), and it is similar to abstract models made upon the scholar's knowledge (Kuhn 1977): it is a way either to strengthen the scholar's knowledge or to test the knowledge put inside the model.

Analyses of the model and necessary activities, such as verification and validation, with ABMs in fact face the problem of the complexity captured by the model. This complexity often implies a vast amount of model outcomes to take into consideration, both quantitative and qualitative, and it makes difficult, or opaque, for the researcher to fully grasp the dynamics happening in the model.

The model, obviously, produces results determined by the micro specification formalized (programmed) by the researcher, but the process of relating outcomes with model components is not straightforward. For instance, there can be concatenations of social mechanisms and overlapping dynamics (i.e., more mechanisms can be at work at the same time). The understanding of model outcomes and the evaluation of the role that different model components play in determining them can thus become problematic. As Di Paolo et al. (2000, p. 6) noted with ABMs, "the behavior of a simulation is not understandable by simple inspection, effort towards explaining the results of a simple simulation must be expended, since there is no guarantee that what goes on in it is going to be obvious."

Researchers using ABMs have thus to implement a systematic sensitivity analysis allowing to complete the map between the generated outcomes and the elements of the micro specification. By inspecting the model (and the modeled system indeed) and by a trial-and-error strategy, it is possible to shed light on the model mechanisms and on interdependences. Analytical results of ABMs often take the form of unintended consequences of researchers' hypotheses, as it happens with any other kind of formal models. In the case of ABMs, the "discovering" of such hidden results, and ultimately the investigation of causality, can require more quantitative analysis than in other modeling tools.

2.2.2   Statistical mechanics, system dynamics, and cellular automata

CSS researchers often use ABMs, but they also use other modeling tools. We shortly describe them to make clear when and why ABMs are preferable.

The first alternative modeling approach to be considered is derived from physics and called "statistical mechanics." It combines probabilistic tools with mechanics, which is the studying of particles subject to a force. The result is a modeling approach that relates the micro level behavior of individual atoms and molecules to the macro properties of the materials made by the atoms and molecules considered, helping in solving problems of thermodynamics.

Because of the presence of a micro level of atoms generating macro properties of materials, statistical mechanics models have been presented as metaphors of societies. Such an approach, in particular, has been widely used in financial studies, giving rise to a discipline called econophysics, where atoms represent investors and "material" properties are market behaviors such as price dynamics (Farmer et al. 2005).

The statistical mechanics approach has been followed even in the field of sociology and social psychology (Latané and Nowak 1997, Nowak et al. 2000), often combined to other tools for simulation purposes.

Sometimes, statistical mechanics models appear as systems of differential or difference equations which represent the dynamic behavior of atoms, and to solve such systems, it is necessary to try many parameters values by using Monte Carlo algorithms. Moreover, because standard statistical mechanics models often do not allow studying complicated structures of interaction, different simulation techniques are used in conjunction with or as extensions of statistical mechanics approaches. An example of this latter kind of integrated approaches is Nowak et al. (2000) where systems are simulated with cellular automata (see the following) to insert a spatial dimension of interaction in the model.

Statistical mechanics is a modeling tool widely accepted in the scientific community, and it is easy to be communicated; however, it raises some problematic issues for CSS in general and behavioral CSS in particular. First, a serious problem is the heterogeneity of individuals. As said in such models atoms or molecules are the metaphoric framework to represent individuals in social systems. They all behave in the same way, being differentiated just by parameters. To make an example, a model could represent a society by means of electrons, which all spin, even if they do it with different speeds and directions.

Heterogeneity at the micro level in statistical mechanics models is strictly bounded, making possible to consider just one kind of behavior per time. In social system, it is common to find a kind of heterogeneity in individuals that is characterized by vastly different behaviors, and this is a feature difficult to model with statistical mechanics.

Furthermore, statistical mechanics models are thought to represent atoms free to move in space according to their forces. Applying such a kind of interaction structure as a metaphor in social systems is a further constrain. In fact, it is not possible to consider different structures of interaction, and the only solution is to integrate statistical mechanics models with other simulation and modeling approaches.

A second kind of framework for modeling complex social systems is "system dynamics," but it is rarely used in CSS. The modeling technique

(Forrester 1961, 1968) has been applied to the social domain in economic and industrial problems, organization and management issues, for urban and public policies, and so forth.

System dynamics exploits differential and difference equations to model influences among model components. Via the equations, even feedback loops among the components are modeled, so to create the systematic nonlinearity that characterizes complex systems.

In other words, while a "standard" dynamic model of differential and difference equations represent the future state of the system depending only on the present state and on exogenous parameters, system dynamics is a tool to simulate systems in which the differential equations that represent dynamics of model components can be interdependent and create feedback loops.

The reason why this simulation approach is less diffused in CSS is that it is a modeling approach bounded to consider a few model components, making difficult to consider many interacting individuals. Moreover, the modeling of interaction structures is quite limited, and it is not possible to consider several kinds of it, such as geographical space. The behavior of model components (i.e., social actors) is bounded to the form of a differential or difference equation, often making the representation of decision heuristics difficult. Finally, system dynamics models lack one of the key features needed for explanatory social mechanisms, that is to say the multilevel structure. In fact, model components are all often on the same level influencing each other, and a differentiation between analytical and ontological levels is not explicitly present, although system dynamics models can be nested. In conclusion, system dynamics can be useful for representing systems made of a few interacting simple components, but it cannot be effectively applied to many kinds of social phenomena.

Thirdly, models for "process simulation" can be considered (such models are also known as discrete event simulations, see Banks et al. 2000). They are models originally developed to simulate manufacturing processes and to better organize them, efficiently managing repositories and production activities.

The focus of these models is on engineering and design problems, and the modeling activity aims at replicating the process of interest as a set of elements which can represent places or activities, with flows of people, materials, or information passing through them. Each model component is modeled as a sort of separate function stopping, transforming, and diverting flows, except for flows that are generated exogenously as an input.

Even such a modeling approach is clearly poor for social systems, which are often not made of individuals executing just one and specialized function as a reaction to the flow of inputs received. Further, this modeling framework lacks the capability to manage realistic social interactions.

Even if with recent software tools model components can be nested to represent different levels of analysis, process simulation lacks many requirements for CSS particularly regarding the interaction structure.

The fourth modeling framework worth considering is "cellular automata." Following Gilbert and Troitzsch (1999), it is possible to say that cellular automata are characterized by the following features:

- There is a regular grid (often of one or two dimensions) composed by homogeneous cells, each one identical to the others (cells, when building models of social phenomena, usually represent individuals).

- Each cell has a state, which can be changed over time (states can represent, if considering social systems, attitudes, or characteristics of individuals).

- Time is discrete, advancing one step per time.

- Each cell, on each time step, modifies its internal state according to a rule that uses two sources of information, the internal state of the cell itself is the previous time step and the internal state of neighboring cells (and such a rule represents individual behavior).

Cellular automata have widely inspired simulations of social processes based on local interaction, such as residential segregation (Schelling 1971) or the already mentioned example of Nowak et al. (2000) where they are used to simulate the spreading of entrepreneurship between neighboring geographical areas.

The constraints imposed by cellular automata are several. The first obvious one is the homogeneity in micro components behavior. The second is the rigid structure of interaction that allows interaction with only the closest neighbors in the spatial dimension. Finally, cellular automata difficultly capture the multidimensionality of social phenomena; thus, the state of cells is unique (i.e., it can represent several variations of only one feature of the individual), and the interaction structure is only one (e.g., not allowing modeling partially overlapped networks such as work and friendship relations). Because of such limits, cellular automata are not the most diffused modeling tool in CSS.

2.3 Tools

We briefly introduce here the main common characteristics of tools for agent-based modeling.

Among the nonproprietary tools available for implementing ABMs, probably the most diffused ones today are NetLogo (Wilensky 1999), Repast (North et al. 2013), and Mason (Luke et al. 2005). NetLogo is particularly addressed at modelers without programming experience. Repast and Mason are more powerful tools providing more possibilities for experienced programmers.

All tools differ because of many aspects, but at the same time, they all share some characteristics, which will probably be essential also for tools that will be made available in the near future. Knowing such characteristics simplifies learning and adoption of different platforms.

First, tools aimed at effectively implementing agent-based modeling need to adopt the object-oriented programming (OOP) approach. Compared with procedural programming, in the context of ABMs, OOP allows immediately establishing a parallel between software objects and model agents. The parallel is useful because of many reasons, and the two most important ones of them are the possibility to treat agents (and software objects) in collections and to establish a further analogy between agents' actions and objects methods. In summary, OOP allows programming the model in a way very similar to how it can be represented mentally or on paper. To say it differently, OOP allows direct and immediate formalization of ABMs in computer programs.

Second, tools are mainly in the form of libraries providing a common set of features to simplify and guide ABMs implementation and execution. Such common support features have identical aims but different names (for a comparison of names of features across tools, see table 2 in Railsback et al. 2006), and they derive from the features provided by the first tool for ABMs ever developed, that is to say Swarm (Minar et al. 1996). They can be categorized into four main themes, which are how to set up a model, how to manage time in the model, how to inspect the internal state of the model, and, finally, how to dynamically show the results of the model.

The set up of an ABM is supported by the provision of a flexible scheme representing main model components, namely, agents, interaction structure, and scheduling system for managing agents' actions. The scheme is usually implemented in software platforms by providing different templates (i.e., a generic agent, commonly used interaction structures, and a scheduling system). The scheme is also useful to keep the model separated from the eventual graphical user interface (GUI) by which the user interacts with the model. This feature supports the modeler also in clearly specifying the values and the ranges of input parameters.

The management of time and agents' actions is crucial in an ABM since in this modeling approach, time is always discrete and model dynamics is led by events. Because agents' actions have to be triggered, tools provide customizable software objects capable to manage the triggering of events and actions. This feature, often denoted as the "scheduling" mechanism of an ABM, can be static and defined ex ante by the modeler as a set of events and actions happening on regular frequencies, or agents throughout the model simulation can dynamically change it.

The third aid provided by tools is the possibility to inspect the internal state of an agent while the simulation is running. It is, in other words, the capability of observing a single agent or a few of them dynamically, accessing all their internal variables. This feature is particularly useful when debugging a model or when inspecting it to understand obscure behavior. This feature is accomplished by providing templates of model agents that implement the capability to be internally observed (or "probed" as in the jargon adopted in Swarm), but it provides results that could be obtained also as customized model outputs (either on file or in the GUI) or by means of a dedicated debugging tool.

Finally, there is the GUI. In fact, for simple ABMs or for preliminary versions of more complicated ABMs, it is often useful to have a GUI not only to manage model execution more intuitively than with a command line but also to present dynamically some of the model results. Tools thus provide a set of software components to quickly prepare a GUI where both the simulation and the values of input parameters can be controlled and where dynamics of most important variables can be plotted over simulation time steps. Moreover, tools provide graphical libraries to dynamically present results over particular kinds of interaction structures, such as social networks or space, either based on GIS maps or on abstract matrixes.

2.4   Critical issues: Uncertainty, model communication

Components of ABMs like agents' behavior and interaction structure can be probabilistic. ABMs often have several nondeterministic components that, put together, determine probabilistic outcomes. Empirical data informs probabilistic components of the model because they define, for each probabilistic variable, the characteristics of the probability distribution from which a random value for that is drawn (e.g., the normal distribution and the values of its mean and standard deviation).

Because computer programs are intrinsically deterministic, probabilistic values are drawn by means of specific algorithms, called pseudorandom numbers generators, that mimic randomness. In order to ensure model replicability for both analytical and validation purposes, modelers usually employ a single instance of a random number generator in ABMs (thus, a single generator draws all needed random numbers), and they keep trace of the random seed used by the generator. The random seed is a parameter that instantiates the generator in a specific state so that anytime that random seed is used, the same sequence of random numbers can be generated.

If researchers use a single random seed, the ABM will explore only one possible value in each probabilistic component and model results would wrongly appear deterministic. The simulation thus requires exploring a large number of values of random seeds, and these random seed values should be recorded and communicated along results and the model.

There is not a right way to identify the number of random seeds to explore since the need of random numbers can seldom be determined ex ante. In fact, even if it is known the number of probabilistic components and the number of simulation steps to consider (which is often not known before), the number of random number draws cannot be precisely valued since some probabilistic components can influence the frequency of others. As an example, consider a model with some agents who probabilistically transmit information to the other agents they interact with and where the number of interacting agents is probabilistic too. To state the problem with different words, before running the simulation, the uncertainty propagation cannot be known by researchers, and propagation can be probabilistic too. The solution is thus to consider a number of random seeds that is large enough to observe enough random draws both in each probabilistic component (e.g., in each agent and in the interaction structure) and in each model outcome.

When the number of random seeds is chosen, and thus also the number of replications of the simulation is determined (each with a different random seed), the correct way to select the random seed values to use is to randomly draw them with another random numbers generator. In particular, such values should be extracted from a uniform distribution bounded by the limit values accepted as random seeds by the random generator exploited in the ABM.

Results of this sort of Monte Carlo simulation by which the ABM must be simulated are in the form of distributions, and thus, they should be reported accordingly. The power of ABM is thus also the capability to account for uncertainty quantification and propagation. In fact, the ABM can be analyzed in order to identify which probabilistic components mostly contribute to uncertainty of results.

Because of verification and replication purposes, and for ensuring model reusability, the computer code by which model are formalized and simulated should be shared and made available to everyone.

However, for model comprehension, that is not sufficient. Before presenting the simulation results and proceeding to show their validation with empirical references, a full description of the model should be inserted in any scientific publication.

Readers in fact cannot understand and evaluate model results if they do not know what is inside the model. In the case of ABMs, that means to fully describe how many agents are considered, how they are characterized (saying, eventually for each one, their individual features), and so on. In particular, agents' behavior and the structure of interaction must be presented because they are the two "leading forces" of the system dynamics.

Models are usually communicated in science by using mathematics and verbal descriptions. In the case of ABMs, the formalization process takes place in a formal language different from mathematics: ABMs are in fact pieces of computer programming code.

Mathematics can be used in communicating ABMs but it is usually incomplete. In fact, mathematics can be used to communicate agents' behavior only if it is expressed as a mathematical function. Agents' behavior in algorithmic forms can seldom be represented in mathematical forms, as well as interaction structures.

The reporting of computer code is a surely complete solution, but it encounters several practical problems in understanding. ABMs are in fact programmed in different programming languages and toolkits, requiring the knowledge of too many programming rules, and it is not assumable that all readers are aware of all technicalities. Moreover, the code should be reported entirely, and that would mean to ask the reader to analyze both relevant and nonrelevant parts of the model (as example of nonrelevant parts of the model code, consider instantiation of variables or loading of external libraries—see the beginning of the appendix).

A solution could be the adoption of a standard and simple programming language (i.e., a metalanguage), easy to understand and to share, even if different from the programming language actually used for simulations. However, today, there are not metalanguages accepted as standards in the scientific community. Moreover, the space needed for reporting the whole model would often exceed publication standards.

The most widespread solution today is to rely on protocols guiding the description of the model, ensuring the presence of all essential elements. Descriptions are mostly verbal but can also contain mathematics and graphs describing algorithms and model structure (e.g., with the Unified Modeling Language—UML). The simplest, informal protocol is just to describe the basic components of ABMs that are characteristics of agents, their behavior, and their interaction structure. A more detailed and today well-accepted protocol is the overview, design concepts, and details protocol, shortly ODD (Grimm et al. 2006, 2010).

3

Observation and explanation in behavioral sciences

Behavioral sciences are focused on the observation of individuals' behavior in order to explain it. In the approach of behavioral CSS, observation and explanation of behavior play two different but complementary roles.

Most of behavioral studies follow an experimental approach that is aimed at observing behavior under different circumstances in order to test hypotheses about the causality of behavior.

In behavioral CSS, observation of behavior is a necessary component of modeling that allows discovering and testing complex causal mechanisms explaining social phenomena. At the same time, in behavioral CSS, the explanation of behavior is not usually required for modeling, and it does not directly explain social phenomena; however, it provides robustness to modeling, to the interpretation of results, and in summary to the validity of explanatory social mechanisms.

To make the point clearer, consider as an example a classical phenomenon of a social standard established through a word-of-mouth dynamics. The

Behavioral Computational Social Science, First Edition. Riccardo Boero.
© 2015 John Wiley & Sons, Ltd. Published 2015 by John Wiley & Sons, Ltd.

behavior of individuals receiving and transmitting personal opinions surely plays a central role in the explanatory mechanism of the social phenomenon. At the same time, the reason why individuals value and share personal opinions is a completely different matter. In particular, the explanation of individual behavior does not directly explain why and how new standards are adopted at the social level, but it allows to define the boundaries of validity of the word-of-mouth explanatory mechanism. If we know, for instance, that the individuals' behavior in this kind of dynamics is motivated by the capability to recognize the authors of opinions and by the personal relationship between individuals, we can bound the validity of our explanatory mechanism to the domains where communication is not anonymous (and at the same time we know that we need to investigate what happens with anonymity).

Although with different aims, the approach of behavioral sciences shares several methods with the approach presented in this book. After introducing the main concepts behind behavioral sciences including those regarding explanatory strategies, we focus this chapter mainly on observation of behavior because of the crucial role it plays in our approach.

The chapter is thus organized as follows. In the first section, we introduce concepts and references to the vast literature and methodology of behavioral sciences. In the second one, we discuss the most diffused methods to observe behavior, and in the third, we briefly introduce the tools that allow implementing those methods. We conclude the chapter by discussing the main critical issues that are commonly found when observing behavior.

3.1   Concepts

In the last decades, there has been an increasing interest on behavioral issues both in psychology and in other social sciences. In psychology, the overcoming of a too arid contraposition between the behaviorist and the cognitive approaches has allowed the adoption of more comprehensive and ambitious research approaches. Similarly, the development of new tools in different fields such as social psychology and neurosciences has enriched the toolset at disposal, and they now allow effectively putting in relationship, from the analytical perspective, behavior with cognitive, neural, and social elements.

In other social sciences, analytical tools and important discoveries developed in psychology have fostered social inquiries based on better models of individual behavior. For instance, in economics, the discoveries of cognitive biases in decision-making has provided new strength and ideas to scholars in the discipline already using experimental methods, and it has lead to the development of the field of behavioral economics.

From a broader perspective, the renovated interest on behavior in social sciences has started to impact on policy issues and on the society at large (e.g., Ciriolo 2011, Dolan et al. 2010, Science and Technology Select Committee 2011).

The approach of behavioral CSS is motivated on similar grounds, and it requires the understanding of the basic concepts and methods of behavioral analysis. The basic analytical strategy adopted in behavioral sciences relies on the comparison of behavior observed in different settings. The comparison allows inference making between the different settings and the resulting behavioral changes. In other words, the focus is on investigating how changes in some independent variables (either modified by researchers or naturally occurring) trigger changes at the behavioral level.

The comparison and, if applicable, the design of different settings can be either blind or driven by theoretical hypotheses. In the first scenario, the analysis is conducted in order to identify patterns, regularities in behavioral responses that suggest the existence of a relationship between variations of the independent variable and of behavior. It is an exploratory activity. Instead, when a theory is available, it is possible to exploit the comparative approach to test the theory or one of its components. In this latter case, the theory (or its component) is falsified by confronting empirical results with the ones theoretically predicted. Testing allows also developing new improved theories because it suggests where and how established theories fail.

In order to conduct a meaningful and robust inference making on the data, data collection (i.e., behavior observation) has to deal with control of nuisance variables. In fact, being able to correctly impute behavioral changes to changes in independent variables is of uttermost importance in order to establish a causal link. Because of this reason, researchers in behavioral sciences prefer, if possible, to use experimental methods conducted in laboratories that can provide more control of nuisance variables.

In particular, the use of experiments provides the capability to control for some environmental conditions and to design sample selection and treatment administration. Further, all these features together allow the replication of data collection. The design of experiments for behavioral sciences is a sub-field of the discipline *per se*, and several references provide comprehensive descriptions of the state of the art (e.g., Kirk 2013). For the sake of this book, we briefly discuss the main components of experimental design as outlined earlier (i.e., control of environmental variables, sample selection and assignment, and treatment design and administration).

An important advantage of experiments conducted in a laboratory is the possibility to control for the external environment and variables. Examples of variables that can be controlled span from physical ones such as environmental temperature to the ones related to information reaching subjects. This latter category is the most relevant for behavioral CSS since knowing which information is available for decision-making is fundamental in order to precisely model behavior in ABMs. By saying that in lab experiments environmental conditions can be controlled, we mean that they can be either recorded or manipulated. This means that in the statistical analysis of results, it is possible to take into consideration environmental conditions (because they have been recorded) and that they can be designed either to stimulate behavioral changes (that is to say defining treatments) or to control for nuisance (e.g., with techniques such as randomized block design—Kirk 2013).

Because most of individual variables neither can be controlled nor manipulated, subjects participating to experiments are usually chosen at random. In fact, experiments cannot be run with large or entire populations of subjects, and thus, it is very important to select samples of subjects avoiding selection biases that would make results not robust or externally invalid. The random assignment of subjects to different treatments is one of the essential characteristics of experiments.

Finally, the design of experiments requires dealing with the organization and administration of treatments. This means firstly to identify the independent variable that needs to be modified in order to stimulate behavioral responses. Second, we need to identify a limited set of possible values or configurations of the independent variable that have to be investigated. For instance, when the independent variable is binary, as it is in simplest drug tests, the configurations are two, one without the independent variable (i.e., without administering the drug) that is usually called the control group or control condition and one with the independent variable (i.e., administering the drug to test) that is called the treatment or the treated condition.

Further, it requires deciding whether to adopt a between or a within approach in treatment administration. The difference between the two is that in the between approach, different groups face different experimental conditions (e.g., the control group is different from the group of subjects that experiences the treatment). In the within approach, it happens that the same sample of subjects face all experimental conditions (e.g., it is observed before taking the drug and after). Because within design implies sequences of experimental conditions since they cannot be administered together at the same time, it is important to design also the sequence of conditions with the aim of minimizing the presence of unaccountable effects. In particular, negative sequential effects appear when the behavior observed in a condition is influenced by interactions and behavior in preceding experimental conditions. For instance, getting back to the example on drug testing, in order to administer first the treatment and then the control condition, it is important to be sure that the drug has completely vanished from subjects' bodies. In order to control for sequential effects, the simplest but not always feasible solution is to design random sequences of experimental conditions across the sample.

Results collected either in experiments or with other methods (see the following section) are then analyzed statistically, most of the times by means of analysis of variance, with the aim of robustly (i.e., controlling for nuisance factors) investigating the causal link between independent variables and behavior.

In behavioral CSS, statistical analyses of behavior aimed at inference making, as discussed in the introduction, are not as essential as observation. In fact, what is needed in behavioral CSS is the capability of modeling behavior, and this does not always imply the need to establish precise causal links between external independent variables and behavior. Most of the times, in fact, the modeling of behavior requires only establishing causal explanations linking the information subjects have at disposal with their decision-making (a more detailed discussion of behavior modeling is presented in Chapter 5).

The modeling of behavior in behavioral CSS, however, requires a precise observation of behavior that includes the information reaching subjects, their decision-making, and possible external factors influencing behavior (i.e., nuisance factors). Methods of behavioral sciences precisely allow observing behavior and those other elements.

3.2   Observation methods

Besides the tools for inference making that are the statistical ones applied to collected observations (e.g., analysis of variance), behavioral sciences use several different approaches to collect behavioral data. In many cases, it is worth considering using a mix of the methods presented in the succeeding text since some methods are preferable for observing specific kinds of variables (for instance, demographic variables are often collected through surveys) or for solving specific critical issues (e.g., external validity). In other words, they often can be considered as complementary methods, not as alternative ones.

3.2.1   Naturalistic observation and case studies

Naturalistic observations are the passive ones made in natural settings. Variables cannot be manipulated, and measurements have to be designed in a way that does not modify natural conditions. In naturalistic observations, researchers cannot always measure all the variables they desire because sometimes the measurement of some variables would modify natural conditions.

For the purposes of the approach presented in this book, naturalistic observations are useful when it is possible to measure information exchanges (either private or public information) and decision-making. Further, it is often useful the possibility to observe demographic information about individuals.

The advantage provided by naturalistic oboservations is twofold. First, since the intrusion of the researcher is kept to a minimum, it is less probable to observe induced behavior (i.e., behavior that individuals adopt induced by the mere presence of the researcher and by their assumption on what the researcher wants to observe, often referred to in the literature as "reactive arrangements"). Second, behavior is observed in real-life situations, and thus, results can easily be extended to them.

Case studies are observations of small samples of populations where the natural conditions are maintained as much as possible but where intrusive measurements and small manipulations can be allowed. They are particularly useful for in-depth investigation of behavior of small real-life situations. Case studies carry on the problem of controlling for nuisance factors, and it is difficult to extend what is observed in case studies to the wider population.

3.2.2   Surveys

Surveys are based on the self-reporting of some information by respondents. They are usually administered to samples of subjects selected according to principles of statistical significance and random blocks design. Responses are collected by means of questionnaires or structured interviews (see the following for the tools to conduct surveys).

With surveys, it is often possible to collect large amount of data, although keeping the sample of respondents representative is often problematic because not all selected individuals respond (and the response rate is low because individuals do not feel confident to share their private information) and because a certain degree of selection biases cannot be avoided. For instance, regarding this latter problem of the presence of selection biases, it is often the case that researchers at the moment of the survey administration do not know how many subjects in their sample have actually been involved in the dynamics of interest, and it is impossible to infer the reasons why nonresponding subjects are such (e.g., whether nonrespondents have nothing to say or they do not want to).

Further, in surveys, people tend to provide responses that are socially acceptable or that respondents think would answer researchers' expectations.

Because of these reasons, surveys are not the preferable tool when observing some specific kinds of situations and behavior, for instance, in the case of illegal activities.

Moreover, surveys should be conducted in temporal proximity to the behavior or event that has to be observed because respondents' memory is limited and its inaccuracy grows as time passes.

Finally, with surveys, it is impossible to observe behavior under manipulated conditions since responses about hypothetical situations are not reliable because of rationalization, social acceptability, and reactive arrangements.

With surveys, it is very difficult to obtain behavioral data along the within approach, that is, the one where behavior is observed for the same subject under different conditions.

Though conducting surveys can be cost-effective, it can allow obtaining good levels of representativeness and external validity, and it allows measuring a large number of variables, much care has to be spent in survey design (De Leeuw et al. 2008) and in the formulation of questions in order to minimize the presence of induced responses and inaccuracy because of misunderstandings, boredom, and tiredness.

3.2.3   Experiments and quasiexperiments

Experiments are usually characterized by three elements. First, they are conducted in artificial (i.e., nonnatural) settings. The artificiality can be limited to a few elements, such as the place of observation (e.g., a classroom in a university instead of a suburban street) or the kind of interaction (e.g., a game presenting a social dilemma instead of the real situation with a similar dilemma), or it can extend to several aspects at once.

Second, in experiments, researchers manipulate independent variables in order to observe behavioral responses. Third, nuisance factors are controlled for with tools such as the artificiality of the place or the random assignment of subjects to treatments.

The second characteristic, which is the presence of manipulations of independent variables, is not always needed for behavioral CSS since, as explained before, the focus is more on observing behavior than on explaining it. Quasiexperiments are experiments where the control of nuisance factors does not include the random assignment of subjects to treatments because it is impossible to do so (e.g., when groups are necessarily made by individuals of different ethnicity).

The artificial environment of experiments is useful not only because of the control of nuisance but also because it simplifies measurement of all relevant independent variables and of behavior. Measurements in fact become components of the artificial setting created by experimenters.

In economics, experiments have a further necessary element concerning the design of the artificial environment: the induced value. In fact, in economic experiments, the focus is on observing behavior that has economic consequences. Since economic consequences are quantifiable in terms of monetary values, it is common practice to "pay" subjects depending on their behavior in order to promote external validity of economic experiments. This kind of payment has not to be confused with show-up fees and other incentives that are only aimed at promoting participation and data collection (payments that can be used also with other methods such as surveys). Further, while in many cases payments in experiments cannot be equal to the absolute monetary values present in the real-world situation that is studied (think, for instance, about an experiment about contributions to a pension fund), their relative values have to be realistic. In other words, when realistic payments are not feasible because of limited budget or because of any other reason, external validity can be promoted thorough a structure of experimental payments based on realistic differences between payments associated with different behaviors.

3.3   Tools

In naturalistic observations, tools for measurements should be the least intrusive possible. Examples of such tools are hidden recording devices and digital traces stored on computer networks or even satellite imaging. Such tools are rarely useful when studying behavior associated with social phenomena, but in these fields, new possibilities are opening every day.

The tools for naturalistic observations more commonly used by social scientists are usually in the form of secondary data made available by public and private data collectors that have to measure such data for reasons of accountability or for the implementation of public policies. Examples of this kind of tools are fiscal data, data about consumption levels provided by utilities, or data about requests for rebates making convenient the installation of solar panels. It is, in different words, a reuse of data already collected for other purposes.

Passing to the tools aimed at administering surveys, their first desideratum is the capability to manage dynamic structures. In fact, in surveys, it is very important to carefully design the questions and the structure of the questionnaire (or of the interview) in order to obtain high levels of attention and high quality in responses. A dynamic structure in a questionnaire and in an interview is the one that dynamically selects the questions to be asked depending on the answers received in preceding questions.

Because dynamic structures are difficult to obtain with paper-and-pencil questionnaires, today, it is preferable to exploit the power of computers and of telecommunication. A first possibility is thus to use the web (i.e., computer-assisted web interviewing—CAWI). Platforms for designing and administering surveys on the web have several interesting features, such as the capability to present texts, images, and videos and to design the graphical interface, but they usually include also very useful tools to recruit, manage, and invite samples of respondents. Among the many open-source projects available today, the one that is probably the most diffused and powerful is LimeSurvey (Schmitz 2012).

When using a computer to conduct interviews by telephone (computer-assisted telephone interviewing—CATI) or in person (computer-assisted personal interviewing—CAPI), a free of charge and very powerful solution is to use Survey Solutions that is the set of survey management tools provided by the World Bank. Among its many advantages, there is the support of mobile teams of interviewers, thanks to the provision of software that runs on tablets and to the possibility to have centralized data management.

Because experiments are usually conducted in artificial environments such as classrooms in universities, tools for experiments usually run on personal computers. However, because the software must allow interaction, and it has to record much information about stimuli and subjects' decision-making, it is fundamental to run experiments on computers that are connected through networks (i.e., computers need to communicate data). Software for experiments has to provide the possibility to show elements like pictures, texts, and videos, and they must collect responses and save results. Further, it is sometimes useful to collect measurements from other devices such as eye trackers. Software platforms for experiments are either based on interfaces running in web browsers (allowing experiments on networked computers of either academic labs or less controlled environments such as the web) or on stand-alone interfaces that need to be preinstalled on the personal computers that will be used by subjects.

Among the tools developed for conducting psychological experiments, there is jsPsych (De Leeuw 2015) that uses JavaScript in web browsers, Tatool (von Bastian et al. 2013) that uses the Java programming language, and OpenSesame (Mathôt et al. 2012) that is based on Python. In economic fields, there are all-purpose programmable tools such as zTree (Fischbacher 2007) and Software Platform for Human Interaction Experiments (SoPHIE—Hendriks 2012) and many other platforms and programming libraries that are designed to conduct experiments based on specific economic situations (e.g., auctions, financial markets, etc.) or on specific kinds of data collection (e.g., MouselabWEB that allows process tracing on web browsers running on personal computers—Willemsen and Johnson 2010).

In general, both surveys and experiments could also be done with paper and pencil, but they would provide lower quality results. In fact, with paper and pencil, it is difficult to implement dynamic structures in surveys and blind interaction in experiments. Further, in both cases, the time requested to respondents/subjects to complete the observation would be longer with negative consequences on attention and response rate.

3.4   Critical issues: Induced responses, external validity, and replicability

A vast literature in behavioral sciences deals with the threats that research methods in general and experimental ones in particular have to face in order to obtain valid inference making. Such threats mainly concern statistical analyses, and both internal and external validity. In the behavioral CSS approach, because the main role in inference making is played by ABMs and because the focus is more on the observation of behavior than on its explanation, critical issues are slightly different.

Internal validity, for instance, is still a concern but from a different perspective. In fact, in our case, the problem is to observe the entire flow of information getting to subjects so that the observation is complete and that all the information used by subjects in decision-making is recorded. If the observation is incomplete, it will be impossible to derive an accurate description of behavior and to model it in ABMs.

Most of the critical issues in observing behavior, however, deal with the problem of external validity. In fact, if experiments are used, there is the problem to evaluate if the artificial environment employed at the lab has any impact on behavior. Similarly, if surveys are used, it is problematic to verify if observations collected over a relatively small sample can be extended to the general population of interest.

Besides, there is the problem of researchers unintentionally inducing specific behavior, especially in experiments. When experiments deal with economic behavior, the problem is even stronger because of the induced value used by researchers to create the artificial environment of economic incentives. In some of these cases, there is a very high risk that subjects tend to behave "too rationally" and on the basis of profit maximization both because the incentive structure and instructions tend to point out that strategy and because the discipline of economics tend to focus much on profit maximization and to present it with a normative flavor.

A certain degree of effects due to the presence of the researcher (acting either as a simple observer or as an experimenter manipulating the environment) is inevitable because the act of behavioral research is a social situation and interaction *per se*. Researchers should pay much attention and spend much effort in trying to minimize such effects; however, there are some strategies that allow evaluating whether such effects make results of observations extendable and acceptable.

In order to evaluate the external validity of observations, the most effective strategy at disposal is replication. Replicating experiments and surveys allows testing whether results can be generalized, and it can suggest the presence of unobserved but relevant variables.

Replicability of observations is thus of uttermost importance when observing behavior. In order to ensure replicability, it is thus important to carefully design the survey/experiment in advance, trying to foresee all its phases and criticalities and the means to record them.

4

Reasons for integration

The development of social sciences has implied a process of scholars' specialization not only in terms of focus on more theoretical or applied research and on different disciplines but also in terms of adopted tools. Such a specialization is not, *per se*, a disadvantage if it does not carries with it the building of uncrossable boundaries.

We propose here an integrative approach that, in a few and imperfect words, joins a modeling paradigm with behavioral empirical analyses. Some advantages of the approach can be foreseen, and a few have already been pointed out and discussed in the literature. In this chapter, we briefly sketch the advantages of our proposal according to different established perspectives.

First, we discuss advantages from the perspective of social scientists used to modeling in general and to ABMs in particular. Second, we discuss the perspective of behavioral social scientists such as experimental and behavioral economists and sociologists. Last, we discuss advantages from a broader

Behavioral Computational Social Science, First Edition. Riccardo Boero.
© 2015 John Wiley & Sons, Ltd. Published 2015 by John Wiley & Sons, Ltd.

disciplinary perspective. Distinctions, however, are loose and our approach aims also at overcoming weakly motivated differences. We thus encourage the reader to go beyond personal perspectives in order to have a broader and more complete view of the advantages of the approach.

4.1 The perspective of agent-based modelers

The methodological debate about the opportunity to have in ABMs accurate descriptions of behavior at the agent level has probably started at the same time of the birth of the tool.

Even completely opposite opinions about this issue have good motivations, and the reason why a single and definitive answer on the issue will never be reached is that ABMs can be used according to different research strategies and methodologies. Consequently, the requirement to have empirically valid behavior in ABMs strictly depends on the model aims.

For instance, ABMs could be used as black boxes, with capabilities of predicting and mimicking some social phenomena. In such a case, perhaps, the modeling of agents' behavior does not need any empirical reference. However, an ABM that is a black box (and even a "good" one because of its predictive accuracy) cannot be used for investigating causality.

In this book (in Chapter 2), we present ABMs as the powerful tool that allows computational social scientists to investigate causality taking into account the intrinsic complexity of social systems. At the basis of the choice of ABMs for our purposes is thus their capability to support the search for explanations in the form of social mechanisms.

Further, different research strategies can prefer using statistical tools if they are aimed at finding significant correlations among unexplored variables or if aimed at testing the accuracy of other kinds of explanations of social phenomena such as covering laws (such as in many applications in econometrics). Similarly, other modeling tools and approaches should be preferred for conducting theoretical investigations under different epistemological hypotheses.

For the approach here presented, it is evident that unrealistic behavior would make the social mechanism unrealistic and invalid. In fact, agents' behavior is an essential component of social mechanisms. Investigating causality means to establish causal relationships between elements in order to explain a phenomenon, and if one of the elements that are considered is known not to be present in the system (that is what unrealism means), there is no point in investigating causality because results will always be flawed by definition.

Besides being a necessary component of ABMs for behavioral CSS, the presence of empirically founded behavior in ABMs provides advantages even under different approaches. We discuss them starting from a twofold disclaimer.

First, the tool integration proposed here is not a standard in the community and, on the contrary, is something that has been presented in only a few contributions and only in very recent years.

Secondly, the most of the early attempts to integrate ABMs with detailed knowledge about individual behavior have been only of a particular kind. They were, in fact, based only upon knowledge derived from economic experiments, and they followed a straightforward procedure by which ABMs were used for the understanding of experimental results (Duffy 2006), avoiding to further integrate the tools in other directions, for instance, by using ABMs for informing and designing new experiments. Following Duffy (2006), the reason was that "as human subject experiments impose more constraints on what a researcher can do than do agent-based modeling simulations, it seems quite natural that agent-based models would be employed to understand laboratory findings and not the other way around."

Nevertheless, in those first attempts of integration, results from experiments were useful for modeling at two different levels. At the aggregated one, they were useful in validating ABM capability of replicating the same systemic dynamics observed in the lab. Following Duffy (2006), they answered research questions as "do artificial adaptive agents achieve the same outcome or convention that human subjects achieve? Is this outcome an equilibrium outcome in some fully rational, optimizing framework or something different?" At the individual level, on the contrary, the focus was on the evaluation of the external validity of the behavioral rules implemented in ABMs, by comparing them with observed behavior.

Those first integrations were applied to two main research themes. First, the large majority of models followed the "zero-intelligence" approach. Second, there were other models that implemented agents' behavior in a more sophisticated and sometime realistic way, with behavioral modeling ranging from simple reactions to stimuli to learning heuristics such as classifier systems and genetic algorithms (Duffy 2006).

In the case of the latter kind of models, the process of model validation often exploited experimental data at both levels (i.e., at the agent and system levels), while in the zero-intelligence approach, the experimental data on which model outcomes were validated was the aggregated one.

The use of ABMs in the zero-intelligence approach, which in economics predates complexity and ABMs, enriches our knowledge about the fact that agents with very low rationality constraints, and sometime even "stupid" agents, are sufficient to reach equilibria in markets and to replicate common aggregated dynamics such as bubbles in financial markets. But if that corroborates the idea that maximizing agents are not always needed, it does not answer the consequent question that is "which agents" can improve the analysis of social phenomena.

Something similar has been also noted by Janssen and Ostrom (2006) recognizing that the initial successes of ABMs were theoretical and abstract (e.g., the famous Schelling's segregation model and the Axelrod's cooperation tournament). But as they correctly state, by using such an approach, "most ABM efforts do not go beyond a 'proof of concept'," and thus, there is a need for the empirical validation and calibration of ABMs, in particular for the micro layer and for agents' behavior.

Other examples of similar applications of ABMs enriched with behavioral evidence come from the field of land-use studies that has been on the forefront of the conjoint use of experiments and ABMs. Such a research field, in fact, has been among the firsts in developing interest on the different methodologies for the empirical foundation of ABMs (Robinson et al. 2007). It is, for instance, the case of the integration with experimental data in Evans et al. (2006), where outcomes from simple experiments with human subjects are compared with ABMs developed by using utility-maximizing agents. In the author's words (Evans et al. 2006),

> The main findings are: (1) landscapes produced by subjects result in greater patchiness and more edge than the utility-maximization agent predicts; (2) there is considerable diversity in the decisions subjects make despite the relatively simple decision-making context; and (3) there is greater deviation of subject revenue from the maximum potential revenue in early rounds of the experiment compared with later rounds, demonstrating the challenge of making optimal decisions with little historical context. The findings demonstrate the value of using non-maximizing agents in agent-based models of land-cover change and the importance of acknowledging actor heterogeneity in land-change systems.

Finally, it is interesting to notice that the interest on the empirical micro foundations of ABMs, and thus on the integration of behavioral analyses, has spread in the ABM community in social sciences since some methodological contributions such as Boero and Squazzoni (2005) and Janssen and Ostrom (2006), but it is also recognized in new fields of application of ABMs, for instance, in biomedical research, which is one of the most recent branches of science in which ABMs have started to spread. It is there clear how "toy models" for abstract reasoning have only a partial value, and further microknowledge is needed for building ABMs (Thorne et al. 2007).

Moreover, elements such as the ever-growing set of available empirical sources (from network statistics to behavioral data) and the often particular kind of sources typical of social sciences (e.g., qualitative analysis—for an example of an effective integration of such knowledge in ABMs, see Tubaro and Casilli 2010) support the idea that to accomplish micro-level empirical salience, researchers using ABMs have to master several data collection and analysis tools and most importantly those of behavioral sciences.

To make the point clearer, consider the interaction structure of agents. If, for the sake of the research question, it can be considered static, there are not criticalities in using the available information in ABMs. The difficulty can be in collecting such a kind of data, but if available, the empirical interaction structure can be modeled in the ABM as observed. If, on the contrary, the time dimension of the model makes a dynamic interaction structure a preferable choice, the issue is to understand agents' behavior in establishing interactions.

Two final doubts remain to be addressed at this point of the discussion. First, is behavioral knowledge essential? Could not we rely only on cognitive theories for informing ABMs, avoiding the burden of behavioral data collection and analysis? Second, should the recent fields of neuroeconomics and of neuroscientific approaches to social sciences be considered as a further useful information source for the micro levels of ABMs?

Answering the first question, we can refer to Richetin et al. (2010) where ABMs are built starting from psychological theories but without a direct reference and validation with micro empirical data. While the example testifies the advantage of considering cognitive models of behavior that include intentionality in comparison with approaches that reduce behavior to a static function, it does not support the usefulness of highly abstract theories of human behavior (such as the ones there implemented coming from social psychology) in explaining social phenomena. On the contrary, that approach is particularly effective for theoretical studies of behavior and cognition.

Similarly, the ABMs that exploit for agent behavior the so-called "rich cognitive" approach (see Chapter 6) require behavioral data and validation to support social explanation. In particular, ABMs built following that approach use agents with explicit and often complicated mechanisms based on different levels of cognitive capabilities. Unfortunately, even in this research tradition, the link with the micro evidence is often left apart; the focus is on cognitive models alone, and thus, the micro–macro explanation is impossible.

For instance, Dignum et al. (2010) present an ABM of trading networks with intermediaries. The ABM is built according to a specific formalism (i.e., MASQ—see Ferber et al. 2009) where agents have a component called "mind" with "the definition of their specific goals (e.g. 'satisfy a need' for a consumer or 'make profit' for a supplier) and the specification of their decision-making strategies (e.g. how to select a supplier? how to compare products? when to make a transaction? etc.)" (Dignum et al. 2010). Further, in those models of agents can be present the evaluation of costs (either of business activities or cognitive ones) and of the levels of satisfaction. The resulting model is particularly useful for theoretical research and helpful for managers who want to better comprehend and change their organization using tools of business process simulation. The lack of micro evidence, however, makes the model not well suited for investigating causality of actual phenomena.

Passing to the question regarding neuroscientific contributions on the studying of behavior in socioeconomic domains, the answer requires a distinction that has been proposed in Ross (2008b) and Davis (2010). According to those contributions, it is possible to divide neuroeconomics in two strongly different approaches and to critically evaluate other neuroscientific contributions to the social domain in a similar way.

The first approach, that can be called "behavioral economics in the scanner" (Ross 2008b), proposes neuroeconomics as "an extension and development of behavioral economics which aims to secure additional new evidence from neuroscientific research for many of the conclusions reached by psychologists and behavioral economists about the behavior of individual economic agents" (Davis 2010). The supporters of this approach are many (e.g., Camerer 2006, 2008, Camerer et al. 2005), and probably most of the contributions appeared so far in this literature follow it.

The second approach opposes the first and has been named by Ross (2008b) "neurocellular economics." According to Davis (2010), it "uses the modeling techniques and mathematics of economics—specifically optimization and equilibrium analysis—to represent the functioning of different parts of the brain without making any assumptions about how neurocellular processes are related to the individual as a whole." Such an approach can easily be generalized to include the contributions of neuroeconomists such as Glimcher (2003) and Neu (2008).

The two approaches completely differ for the research agenda, the experimental design and the collected data. The latter approach, in particular, is focused in exploring the possibility to model the brain at the neural level and, as such and for now, cannot yet be effectively integrated in behavioral CSS.

4.2   The perspective of behavioral social scientists

Critics of experiments in social sciences often point out that they lack realism and generalizability (Falk and Heckman 2009). Opinion may diverge about the legitimacy of such a critique, and, surely, it is also worth remembering how much experimental research in social sciences is heterogeneous because of methodology and interests. However, the integration of experiments and behavioral analyses with ABMs also aims at improving realism and generalizability of experimental evidence by means of new possibilities. In fact, with ABMs, experimenters can solve common critical issues of experiments along three main dimensions: scalability, time, and interaction.

Scalability refers to the fact that experiments are expensive, and the number of subjects is bounded, and often, it is not representative of the universe of reference. The replication of experiments for improving confidence about the obtained outcome is expensive too. Furthermore, experimenters often do not know if observations are influenced by the (small) scale that has to be adopted.

ABMs can solve many of the just mentioned criticalities. In fact, they allow the replication of experimental dynamics all the times that are needed. Further, they allow changing the composition of agents in the model (i.e., modifying sample composition) according to different distributions of specific individual variables in order to obtain a higher representativeness of the universe of reference. Finally, allowing scholars easily exploring the model parameters by changing the number of agents, they can be used to verify if scale matters.

To make the point clearer, imagine of studying a social dilemma having conducted an experiment about it with some dozens subjects recruited at the university. In order to be surer about the outcomes, it would be better to replicate the experiment and double, at least, the number of subjects but that can be too expensive or impossible because of the bounded pool of students available for experiments. An ABM allows replicating the experiment with artificial agents that, if correctly modeled, can represent additional human subjects. Furthermore, by knowing some specific features of students comparable to those in the generic population, for instance, an hypothetical inverse relationship between age and willingness to cooperate, experimental results can be projected on a more balanced population increasing external validity. Finally, the impact of group size and sample size can be investigated with ABMs almost at no cost (see the example presented in Chapter 8).

Obviously, changing crucial parameters of ABMs built upon experimental knowledge could mean a loss in robustness and validity of the behavioral model; thus, it is often preferable to verify some of the model suggestions via new experiments. In this sense, ABMs support the design of experiments (Duffy 2001).

The second dimension on which the proposed approach seems to be most promising is time. In fact, in experiments, the repetition of tasks is bounded by costs, subjects' tiredness, and attention. Moreover, in social phenomena, agents' behavior affects dynamics which have different timescales.

For instance, saving and consumption decisions are made by households on a daily basis, but capital accumulation changing the aggregate production level happens, or to be clearer "significantly manifests its changes," on a longer timescale. ABMs allow extending the number of repetitions of tasks toward a horizon of time longer than any experiment can do. Furthermore, while experiments need to focus on a specific dynamic and to control the others by keeping them static to reduce nuisance, ABMs can allow considering all the relevant dynamics at work together, even if they have a different timescale.

To make the point clearer, two examples can be referred to. Firstly, Hailu and Schilizzi (2004) analyzed biodiversity conservation in private lands. Their work focuses on the efficiency of allocation schemes, and it demonstrates that the continuous auction which has been pointed out in the relevant literature as the most efficient one cannot guarantee any superior performance if subjects are capable of learning from past experiences. ABMs equipped with learning agents (according to some plausible hypotheses from the perspective of cognitive capabilities of decision-makers) demonstrate many shortfalls of continuous auction compared with other solutions such as regulations and negotiated contracts. Those fallacies of continuous auction could not have been pointed out by means of game theoretical approaches or experiments due to the boundaries on rationality assumptions (in game theoretical models) and on the number of repeated auctions (in economic experiments).

Secondly, it is possible to consider some of the many available contributions exploiting ABMs and role-playing games (RPGs) to observe behavior concerning natural resource management. RPGs are quasiexperiments particularly focused on interaction in decision-making. They "have been used to understand or support [...] collective decision processes as well as to train stakeholders involved in them," and they "make misunderstandings among stakeholders explicit by splitting the decision process among several decision centers" (Barreteau and Abrami 2007). Though similar to experiments, the loosely controlled environment of RPGs and their "naturalistic" sample selection make them difficult to replicate.

The example, presented in Barreteau and Abrami (2007), focuses on the different timescales happening in the different modeled processes. The field of application, as mentioned, is natural resource management and in particular water-sharing dynamics. Such a critical management involves several behaviors characterized by different timescales identified as follows (Barreteau and Abrami 2007):

> There are basically at least four timescales to be considered:
> 1. The operational timescale deals with the practices of resource use, which are typically at the day timescale or even shorter.
> 2. The strategy timescale deals with the design of strategies guiding these practices, which are usually seasonally based. These strategies constitute the basis for the determination of choices at the operational timescale level.
> 3. The constitutional timescale deals with investments or collective rule design, which has a longer characteristic time of a few years. This is the time of the constitutional choices that frame the design of strategies at the strategy timescale level.
> 4. The resource timescale deals with resource dynamics, which is not hierarchically linked to others but, for some resources such as forestry, might even reach several dozens of years.

The authors implemented a conjoint "model" (an RPG with an underlying ABM) that, because of the possibility to consider different timescales, allows stakeholders in understanding water shortages happening on a specific river basin. Without such an integrated solution, it would have been impossible to understand the impacts of daily decisions on phenomena occurring at different timescales.

The third dimension in which ABMs can improve the analytic and predictive power of experiments concerns interaction structures and dynamics. This dimension abundantly overlaps with the preceding ones, as it will be made clear in presenting some examples, but the main problem that is on focus here is the external validity of interaction structures implemented in experiments. In fact, in order to control the environment, to reduce costs, and to make the experiment easily replicable, interaction structures in experiments are often very simple, static, and unrealistic. On the contrary, actual social agents are embedded in a social environment that they shape and by which they are much influenced. Social interaction here recalls again the multilevel nature of causality of social phenomena we discussed in Chapter 2.

Experiments can only sketch such a complex multilevel interaction. However, the presence of multilevel causality strongly impacts agents' behavior because the environment creates opportunities and boundaries to human action and the social environment influences human behavior even more deeply.

Referring to the work of Bosse et al. (2009) where ABMs about criminal behavior are discussed, those models are introduced as follows:

> These models may incorporate behavioral agent models from an external perspective, as well as models for internal dynamics. Behavior models from an external perspective may involve rather complex temporal relationships between (1) external factors in the criminal's social context that occur as stimuli and (2) their actions, mediated by their characteristics. Models of internal dynamics usually can be expressed by direct temporal/causal relationships between stimuli and internal states, between different internal states, and between internal states and actions. The internal states may involve, for example, cognitive, decision, reasoning, normative/ ethical, attentional, emotional and personality aspects. In addition, they may involve underlying biological aspects, such as specific types of brain deviations, levels of hormones or neurotransmitters. Examples of aspects related to the wider social context are the observed level of social or organized security control, expectations about acceptance of certain actions within society or within a peer group, and social dynamics within groups.

Issues related to the interaction structure (e.g., the influence of the social context) are thus coupled with those of different timescales (e.g., changes in neurotransmitters, individual behavior, and social context). In general, whenever several levels are considered, several timescales should be considered as well, and vice versa.

Getting back to the example about criminality, Bosse et al. (2009) assumed three types of violent criminals and for each one of them identified the cognitive and behavioral aspects as well as the social ones. Combining such information with biological information (for instance, the impact of alcohol and drugs on psychopaths) and the modeling of an environment with spatial interaction and agents behaving socially (e.g., modeled upon the BDI scheme—see Chapter 6), the different kinds of violent criminals are simulated in different settings, and conclusions can be reached about violent behavior in different environments. In other words, complex but satisfactory explanations of actual social phenomena are rarely obtainable if a reductionist approach is adopted on such an important element as social interaction.

An example with closer applications to economics is the one presented in Dubé et al. (2008). It deals with compulsive choices, where it is evident how much social context and interaction shape the action set of individuals and their impulsivity (without falling on the opposite extreme of reducing individual choices to the environment). Following Dubé et al. (2008), humans often behave in a goal-directed manner, thanks to their will power and self-control, but

> It is possible, however, that a person's ability to engage in such executive control processes is itself vulnerable to environmental influences. Many imperatives of everyday personal, familial, and professional life in modern society render particularly fierce the competition for the cognitive resources necessary for constraining impulsive responses. Thus, the biological influences on motivated behavior call for models of individual choice that are biologically plausible and consider the environments in which choice occurs.

The example points out how much an integrated experimental and modeling approach could give to scientific understanding and experiment-based approaches allowing in fact to consider both the individual and the environmental dynamics involving human cognitive functions for social and economic decisions. Further, the reference to the biological and neuronal side of decision-making is supported by the fact that one of the strongest drivers of food choice and eating is the rewarding nature of food defined in hedonic brain systems. Neuroimaging studies of the brain response to food imagery indicate that the human brain has automatic approach responses to food by comparison to nonfood objects (e.g., Killgore et al. 2003).

Knowing that the reward system is so much influenced by cultural (e.g., healthiness), social (e.g., status symbols), and economic issues (e.g., marketing, pricing, etc.), it is even more evident the need for an integrated approach. This approach starts from situated and controlled experimental and empirical evidence about behavior, and eventually on its underlying cognitive and neural mechanisms, and it explicitly deals with the bidirectional link of explanatory social mechanisms, allowing improving our knowledge about both the individual and the social system.

Continuing with Dubé et al. (2008),

> The complex and dynamic relationships between biology, behavior, and environment affecting motivated choice at the individual level must inform aggregate models of choice. It is at this level that consumer demand drives decisions by businesses and other society systems that shape the balance of eating opportunities and constraints in the environment in which individual choice occurs [...]

The collection of biomarker information (including brain activity) and food consumption data on aggregate as well as individual levels in both their social, spatial, and temporal contexts will become increasingly commonplace and will play a crucial role in linking neurobiological and choice behavior research. Dynamic and spatial structural models are needed to handle these new data opportunities. We expect that agent-based complex systems models will play a crucial role in providing the needed analytical platform for examining the relationship between consumer demand and choices made by businesses and other agents involved such as policymakers.

Similarly, tool integration can interest social neuroscientists: computational models of the brain (for a recent review, see Friston and Dolan 2009), which are crucial in improving the understanding and the design of experiments, can be improved by introducing the capabilities of models such as ABMs and allowing the full consideration of the social environment in which individuals live.

Finally, further examples of integration that can interest experimental social scientists concern group behavior and social dilemmas. For instance, in Goldstone et al. (2008), group behavior is experimentally studied by an Internet experiment. Then, the outcomes of the experiment are studied through an ABM aimed at understanding and predicting the nexus between the "beliefs, goals, and cognitive capacities of individuals and patterns of behavior at the group level," coping in particular with realistically shaped social networks. Finally, social dilemmas studied in laboratory experiments can be extended with ABMs, as in the example on the voluntary provision of public goods presented in Chapter 8 or as in the dictator games studied in Ebenhoh and Pahl-Wostl (2008).

4.3   The perspective of social sciences in general

From the broader perspective of social sciences, the first important reason for integrating the tools we discuss here (i.e., behavioral analyses and ABMs) is related to the multilevel nature of socioeconomic phenomena presented in Chapter 2. The multilevel organization of social systems cannot be reduced, and that is one of the main reasons why social sciences are often considered particularly difficult to study.

Knowing that behavior and cognitive processes are influenced by group level emergent behavior and constraints but that they also are the cause of such macro phenomena, it is worth trying to enrich individual behavioral analyses with the social dimension and to enroot models of social-level phenomena with rich empirical knowledge about individual behavior. The expected outcome of such an attempt is, obviously, a better knowledge of both the society and the social agent who, as we know, is not a castaway on a desert island.

Further, the integration can bear advantages not limited to behavioral studies and social phenomena. On the contrary, it can support interests in cognitive processes and structures (Goldstone and Janssen 2005) and in the brain. In particular, effectively coping with systemic complexity requires acknowledging that a relevant part of the nonreducible complexity of such systems is in the agents' brains. For instance, since the discovery of mirror neurons, affective behavior has gained attention in the explanation of economic choices. When dynamic social interactions are considered along the affective domain, explanation of socioeconomic phenomena has to take into consideration the mutual interdependence between the two, being affect a mechanism often reacting to and defining interactions (Ross 2008a).

Finally, even without considering the contribution in answering research questions of causality in social sciences, integrating traditionally different research tools and approaches enriches individual traditions. In fact, several examples in the literature (e.g., Evans et al. 2006, Heckbert et al. 2010, Janssen et al. 2009) point out the possibility to use different tools iteratively and complementarily, depending on the analytical requirements and contributions specific of each tool. Social scientists equipped with a richer toolset can effectively overcome "false" disciplinary boundaries set by limitations of approaches and tools, and they can contribute more effectively to a cumulative process of discovery.

Part II

BEHAVIORAL COMPUTATIONAL SOCIAL SCIENCE IN PRACTICE

5

Behavioral agents

The capability of modeling individuals' behavior is at the core of the integration of the behavioral and the CSS approaches. Modeling behavior can be done following two rather different methodological approaches. The first focuses on behavior only, it leaves the burden of explanation of behavior outside the model, and it is the focus of this chapter. The second approach employs more sophisticated models of agents that include cognitive functions and structures, it aims at explaining in ABMs both behavior and social phenomena, and it is the topic of the chapter that follows this one.

We call "behavioral agents" the models of agents where the explanation of behavior is left outside of the model. Methodologically speaking, behavioral agents show a few important characteristics worth being discussed in this introduction. First, they are not black boxes, and thus the causal relationships leading to agents' decisions must be clear. Second, the methodological choice of using behavioral agents does not carry on any reductionism. Third, they are models deeply enrooted in empirical evidence.

Behavioral Computational Social Science, First Edition. Riccardo Boero.
© 2015 John Wiley & Sons, Ltd. Published 2015 by John Wiley & Sons, Ltd.

The fact of leaving explanation outside the model does not mean that behavioral agents are black boxes. On the contrary, they have to be white boxes, completely transparent, allowing the understanding of the reasons for agents' decisions and the investigation of the relationship between behavior and social interactions. To make the point clearer, we refer to the schema presented in Figure 5.1.

Figure 5.1 points out that agents' behavior is a "map" that transforms the information agents collect in the social environment into actions. The fact that behavior is not a black box means that this map is made by a set of causal relations that explain how a set of inputs determines a specific output. In other words, with behavioral agents what is included in the model is how agents behave.

The reasons that determine how agents behave (i.e., why agents behavior shows that particular causal structure) are not considered in the model. The fact of adopting this "behaviorist" approach does not necessarily imply reductionism either. In fact, we are always dealing with modeling choices that are aimed at explaining social phenomena. Our approach underlines that behavior and social interaction are the two necessary elements for explaining social phenomena with agent-based models. The reasons for specific behavior can either be left out of the model (for reasons of analytical effectiveness and parsimony) or investigated with more sophisticated agents and other tools if needed (see Chapter 6).

Finally, behavioral agents, besides being nonreductionist white boxes, are characterized by a strong empirical salience. In fact, if behavior is one of the two core components of modeling and explanation, its empirical validation becomes necessary to have valid (and falsifiable) explanations of social phenomena. The need for behavioral validation obviously poses this approach in opposition both to the "as-if" principle justifying Olympic rationality in agents and to low rationality approaches much adopted in the agent-based modeling literature (e.g., following the "keep it simple, stupid"—KISS—principle).

The need for empirical salience in behavioral agents requires researchers adopting behavioral CSS to pay a specific attention in data collection. This

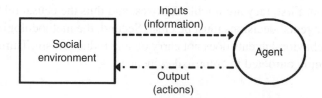

Figure 5.1   The interface between agents (behavior) and the social environment.

issue is the focus of the first section of this chapter. The second section focuses on the tools available when generic models of agents' behavior are available, but they need to be calibrated before being integrated in the ABM. The third section instead focuses on the cases where the models at disposal already include calibrated parameters, and the aim of the modeler is to measure the salience of such models in the empirical sample at disposal, classifying observed behavior and ultimately allowing modeling behavioral agents. The chapter ends by discussing the critical issues that are usually faced in developing behavioral agents.

5.1 Measurement scales of data

The methods for observing behavior and collecting behavioral data have been introduced in Chapter 3. Reminding that such data can be collected in different settings (e.g., in an experiment conducted in a laboratory, in the field through a survey, etc.) and that measurements problems require particular attention, we focus here on the consequences on agents' modeling of the characteristics of behavioral data.

Figure 5.1 introduces the kind of data that is required to model an agent. We need to observe the set of information available to the agent at the moment of the decision (i.e., the input) and the decision taken (i.e., the output). An accurate measurement of this information is what allows calibration and validation of agents' models.

From the perspective of modeling behavioral agents, as it will be made clear in the following two sections of this chapter, two main possibilities are at disposal. The first one is to have a "loose" model of agent behavior that needs to be calibrated (at the individual level or at the population one—see in the following text) on the data. The second possibility is to have at disposal a set of fully predetermined models of behavior and to use the data in order to classify collected observations in terms of consistence with theoretical models.

When models have to be calibrated, the main tool at disposal is regression analysis. When regression analysis is needed, the kind of measurement scale of observed variables most of the time determines the choice of the particular regression model and estimation technique that have to be adopted.

Because of this reason, and because often the kind of observed variable is an artifact due to the measurement adopted (i.e., it is not an intrinsic characteristic of the variable but the result of an inevitably imprecise measurement), it is important to take into consideration different kinds of measurement scales of variables in order to integrate modeling needs in the design of data collection.

In fact, whenever researchers have to collect first-hand data, the design of data collection can determine the kind of variable that will be observed through the choice of the measurement scale that is adopted. Similarly, even when researchers use second-hand data collected by others and thus the design of data collection cannot be modified ex post, often data types (and measurement scales) can be transformed as it will be discussed in the following section and in Chapter 8.

Most importantly, relevant impacts of data types are mainly due to output variables. In other words, the scale and measurement of input variables can have impacts on the regression analysis, but this is not always the case. On the contrary, types of output variables always influence the methods to adopt in modeling behavioral agents, and thus more attention has to be focused on those variables.

First, variables can be cardinal. With cardinal variables order is meaningful (e.g., $2 > 1$), they can be multiplied by a scalar, and the magnitude of their differences is meaningful too (e.g., $4 - 2 = 9 - 7$). Second, there are ordinal variables, for which the order is meaningful (e.g., $low < high$) but on which mathematical operations cannot be performed. Final, there are categorical (or nominal) variables that are variables without numerical values (e.g., occupation, gender, etc.).

The choice of the type of variable and of its measurement scale is often determined by good practices in data collection, by the limits of data collection tools and by the characteristics themselves of the target to observe. In behavioral CSS, further, there is the need to use such data later in modeling agents' behavior.

Moreover, further peculiarities of different data types and measurements have to be considered in the modeling of agents. For instance, a cardinal variable can be continuous in reality but if collected with a survey has limits in precision, and thus it becomes discretized. The modeling of behavior in agents has to face this problem, either allowing the model to be more precise than data collection following the assumption that the measurement error has no significant impact on the data or replicating into the model a similar measurement error (i.e., approximating agents' choices from the continuous space to the discrete one). Similarly, it can happen that measurements are done on a bounded scale for which some of the data is excluded or transformed. For such cases of artifactual limits in the data interval, the researcher has first to adopt the correct regression strategy for either truncated or censored data and then to decide how to deal with limits to behavior in the model (e.g., replicating similar intervals and data truncation in the model).

5.2   Model calibration

Models of agents' behavior are made by causal relationships that determine which output is chosen depending on the input received. The structure of such models, which is the structure of causality flows, can be derived from the vast amount of theoretical and empirical studies or it can be just the representation of relationships inferred on the data at disposal.

Sometimes, models for agents' behavior are loosely defined, meaning that relevant causal relationships are identified but not quantified. For instance, previous behavioral studies can suggest that a specific input significantly impacts the choice of the output but the precise quantification of such an impact is not available. When this happens, it is needed to calibrate the model on the behavioral data at disposal in order to discover the "best" value for the unknown parameter (i.e., the value of the impact of that input on the output), being the "best" parameter value the one that describes the data better.

For the calibration of behavioral models, there are several tools at disposal. A comprehensive list of them would be impossible since the continuous development of tools for data analysis. In the remaining of this section, we present the most important of them available today, following an increasing order of complexity in the "shape" of the behavioral model.

In fact, not all behavioral models are the same from the viewpoint of the structure of the causal relationships modeled. For instance, there can be models where the decision variable is unique (i.e., agents choose only one action) or models where there are many decision variables to be determined at the same time (e.g., partner selection and action to be executed with the partner). There can be models that are simple additive functions (e.g., $y=a+b+c$) and models that are intricate decision trees. Further, there are models and calibration tools that work for cases when a single behavioral model can represent the entire population considered and others that are preferable when models have to be calibrated individually (i.e., separately for each individual).

5.2.1   Single decision variable and simple decision function

When the decision variable is single, the simplest shape of the causal model of behavior is a mathematical function.

If the decision variable is y (i.e., the output of Figure 5.1) and it is cardinal, the generic shape of the behavioral model is $y=f(x)$ where x are the explanatory variables that is to say the information available (or the input in Figure 5.1). Whatever the specific shape of the function, the best tool at disposal in this case is a linear regression (Hayashi 2000, Verbeek 2012). In fact, different functional forms can be estimated with linear regression or the function can be linearized just for regression aims, and then parameters can be retransformed in their nonlinear expression. Further, the usage of linear regression does not imply that the behavioral model has to be linear because it can consider quadratic values of explanatory variables and other powers of variables.

Considering the simplest case, the behavioral model would be $y=\beta X+\varepsilon$ where β are the values of the parameters calibrated (i.e., estimated) on data. The importance of using a regression is in ε, that is to say the error term. In fact, by using a regression, we implicitly acknowledge that the data at disposal can have measurement errors and thus that the data is imprecise. Regression analysis allows us the calibration of the behavioral model under the assumption that the error is present but "well" distributed.

The simple linear regression model can be used for calibration over the entire population or for single individuals. In terms of analytical power of the resulting ABM, the best would be to consider all the richness derived from the heterogeneity in agents and thus to conduct individual-level calibrations, but panel data is needed with multiple observations over time. However, often, the behavioral data at disposal is not sufficient to obtain robust estimates at the individual level, and thus a population-level calibration could be preferable.

In other cases, it is convenient to consider individual differences not by estimating single and separated models, but only modifying parts of the population-level model. For instance, it is possible to estimate a model like $y=\beta X+f+\varepsilon$ where the population-level parameters β are estimated along f that is a specific parameter (or effect) different for each individual. The model, called a fixed effects regression (Baltagi 2013), allows the calibration at the population level with corrections for each individual.

Similarly, it is possible to use random coefficients regression models (Swamy 1970). In these latter models, the calibration is apparently done as in the simple linear regression applied at the population level (i.e., $y=\beta X+\varepsilon$). However, in this case, β values are different for each individual and made by a specific and individual component summed to the average value in the population. In other terms, in these models, the researcher acknowledges that individuals on average behave the same but without renouncing to individual heterogeneity.

In other cases, heterogeneity of behavior in the population is not much due to individual differences but to other characteristics. In these cases, it is possible to use random effects models (Baltagi 2013) or the so-called mixed models (Searle et al. 1992) that allow considering hierarchies in the population, as, for instance, when individuals are students and heterogeneity is mostly due to attending a specific school in a specific school district.

In other cases, different statistical tools are required to cope with specific characteristics of the data at disposal. For instance, with cardinal discrete decision values and when using single-level functions as behavioral models, it is preferable to use "count models" (Cameron and Trivedi 2013), and specific ones when the data is statistically over dispersed (i.e., when the standard deviation of the decision variable is greater than its mean).

Similarly, when using behavioral data collected in laboratory experiments or surveys, it is often worth referring to particular kinds of linear regression techniques such as censored (Tobin 1958) or truncated regression models depending on whether the data is censored or truncated.

On the contrary, when the single decision variable is ordered, there are other tools at disposal to calibrate single-level behavioral functions. In fact, the use of linear regression in these contexts would be an error since the output is nonlinear.

To overcome the nonlinearity problem, the solution is to assume a nonlinear function and to estimate its parameters. The tools at disposal are thus the ordered probit regression and the ordered logit regression, where the parameters estimated refer to explanatory variables transformed by a probit link function and a logistic function, respectively.

When the single decision variable is categorical, a single-level behavioral function can be calibrated using the multinomial probit regression and the multinomial logit regression. Besides the fact that the decision variable is categorical, these regression models employ functions similar to the ones mentioned in the preceding paragraph.

In summary, when the decision variable is single and the behavioral function is simple and made by a single level (i.e., by a single mathematical function), calibration can rely on statistical regressions that allow taking into account the imprecision of data and studying both population-level models and models with more heterogeneity within the population (if the data is sufficient). The tools at disposal and the conditions under which they should be used are synthetically presented in Table 5.1.

5.2.2   Multiple decision variables and multilevel decision trees

When behavior determines the choice of more than one variable at once, a possibility is to conduct a separate calibration for each decision variable. In other words, this choice means to build a different behavioral model for each decision variable and to consider such models completely independent.

If this latter assumption about the independence of decision processes is weak, there are a couple of solutions at disposal. First, if all decision variables are

Table 5.1   Tools for the calibration of single-level behavioral models with a single decision variable.

Kind of decision variable	Level of application of the behavioral model	Regression models
Cardinal	Individual or population	Linear
		Censored
		Truncated
		Count models
	Population with individual differences	Fixed effects
		Random coefficients
	Population with group differences	Random effects
		Mixed models
Ordinal	Individual or population	Ordered probit
		Ordered logit
Categorical	Individual or population	Multinomial probit
		Multinomial logit

cardinal, it is possible to use the so-called seemingly unrelated regression model (Zellner 1962). It allows estimating parameters over a set of functions similar to the ones described earlier for simple linear regressions but with different dependent (decision) variables in each function and even with different explanatory variables (i.e., inputs) in each equation. Further, equation can also be nested, having a decision variable entering the inputs of another decision variable. Using this technique instead of separated regressions on single functions implies that error terms can be correlated across equations, and thus the assumption on independence between behavioral decisions is not required anymore.

When multiple decision variables are noncardinal (either ordinal or categorical) or when they are of different types (of any kind, for instance, one cardinal, two ordinal, and four categorical), a tool that can be used is structural equation modeling (SEM—Acock 2013). SEM provides also many more advantages from the calibration viewpoint. In fact, it is particularly useful for the calibration of nested multilevel behavioral models for which causality paths are well known from the theory, and it also allows considering "latent" variables. The presence of latent variables, that is to say variables relevant for behavioral decision model that have not been observed for any reason, is very common in particular when using behavioral models strongly influenced by cognitive theories.

Finally, there are tools for the calibration of nested multilevel decision models that are made by trees of conditional statements (i.e., if a condition is true, then do something, else do something else). With these models, the most common option is renouncing to use regression tools that allow considering errors in data measurement and to exploit tools that partition information (data) to infer decision trees that plausibly describe the process that has lead to the observed behavior.

Techniques for decision trees induction (Quinlan 1986) have been mainly developed in the field of machine learning, and, from the behavioral perspective, they produce results comparable to and that can be approximated by (multinomial) logit regressions (Joos et al. 1998).

Further, the calibration approach based on the partitioning of the data is also the one to adopt when modeling behavior with Markov chains (Kemeny and Snell 1960). Markov chains can be considered multilevel decision trees with a single decision variable that has to be noncardinal. Further, the decision variable is also the only input at disposal, and it is usually called the agent "state." In these models, in fact, agents probabilistically change state only depending on the state where they are (i.e., they are memoryless and the only information considered is the contemporary state). The calibration of Markov chains is done by partitioning the data by different states and then by studying the distribution of state changes.

Finally, when considering multilevel decision trees with conditional statements and for any type of decision variable, it is possible to use genetic programming (GP—Koza 1992). GP, also known as "symbolic regression" though it does not explicitly consider the presence of measurement errors in the data, is aimed at inferring the shape of a decision tree that represents the data and at calibrating it at the same time. Researchers using GP in fact do not have to specify the shape of the behavioral model ex ante, but instead they are required to specify the space of possible decision trees by specifying the trees alphabet (i.e., the set of conditional statements, input and output variables and mathematical functions that can be used in decision-making). GP infers the decision tree that better describes the data by means of a heuristic that explores the space of solution similarly to genetic algorithms (see Chapters 6 and 8).

The tools described here and their characteristics are summed up in Table 5.2.

5.3   Model classification

A completely different approach can be followed when precise models describing human behavior are already at disposal. Under this approach, the empirical salience of behavioral modeling is guaranteed by examining the behavioral data and by classifying observations and individuals.

Table 5.2   Tools for the calibration of multilevel behavioral models with a single or multiple decision variables.

Kind of multilevel behavioral model	Kind of decision variables	Tools
Nonconditional	Multiple Independent Cardinal	Linear regression
	Multiple Interdependent Cardinal	Seemingly unrelated regression
	Multiple Interdependent Cardinal, ordinal, categorical	Structural equation modeling
	Single Ordinal, categorical	Markov chains
Conditional	Single Cardinal, ordinal, categorical	Decision tree induction
	Multiple Cardinal, ordinal, categorical	Genetic programming

In other words, the approach requires starting from a set of behavioral models (i.e., more than one) and from a database of behavioral observations. Considered behavioral models still need to be white boxes describing causal relationships leading to decision-making, as previously discussed. The following phase is the analysis of the database in order to classify observations into considered models of behavior. By doing this, it is possible to calibrate the ABMs with the observed behavior. From the viewpoint of the behavioral CSS approach and of the analytical power of the resulting ABM, the key issues are here to start from good behavioral models and good data, and then to correctly apply the classification algorithm.

The simplest classification algorithm, particularly useful for the case when decision variables are categorical, considers a binary classification function. Data is organized as aforementioned, that is to say with a set of inputs reaching the agent and one or more decision variables (outputs). Given a set of behavioral models, we compute the predicted behavior in each observation given the inputs and according to each behavioral model. The classification function then simply verifies if the predicted value of the decision variable is equal to the one observed, and it thus assigns 1 if that is the case and 0 otherwise. The procedure is repeated for each observation in the dataset, and results are summed up for each behavioral model and by individual (or by event, depending on modeling aims). If the dataset is unbalanced, meaning that not all behavioral models can be used for prediction the same number of times, results are normalized by computing the mean value. Finally, for each individual is chosen the behavioral model performing better, which is the one providing the highest result.

When decision variables are ordinal and categorical, the classification is based on the same principle of minimization of the error due to predicted values, but in this case, the function is not binary, but the error is computed as root-mean-square errors between predicted and observed values. Similarly, if classification data is unbalanced, results can be normalized in order to allow comparison between behavioral models.

Finally, if decision variables are multiple and of different type, a specific classification function has to be decided in order to weight predicting accuracy across different variables.

If the data is large enough, classification functions allow using statistical tests to verify if the behavioral model providing better results is also significantly better than others in statistical terms. However and at the same time, the procedure does not guarantee avoiding the possibility of multiple solutions for some individuals, that is to say to have multiple behavioral models providing the same predicting accuracy. In these cases, three solutions can be suggested. First, it is possible to try to refine the classification function in order to increase the award given when predictions are accurate. Second, it is possible to refine the behavioral models considered in order to better differentiate and describe the data. Last, depending on the behavioral models that are considered, it is sometimes possible to choose between models equally performing on the data because of theoretical reasons or because of previous studies (see the example in Chapter 8).

In summary, the basic idea behind classification is to compare hypothetical models depending on their capability to describe the actual data that has been observed (i.e., minimization of prediction error). In order to classify the data, it is important to use the same classification function for each behavioral model considered and to normalize results whenever observations are not evenly distributed.

In conclusion, precise and complete behavioral models that can be considered for classification are many and very much context dependent. Many, very useful and appropriate, can be found in the heuristics literature, in particular among fast and frugal heuristics (Gigerenzer and Selten 2001) and social ones (e.g., tit for tat, Newell 2005, Soll and Larrick 2009, etc.). Heuristics are in this context particularly useful because of their solid grounding in the theoretical and empirical literature but also because they are easily transformable in complete and precise behavioral models within ABMs.

5.4   Critical issues: Validation, uncertainty modeling

Most of the criticalities connected to the modeling of agents behavior are not dependent on the techniques presented in this chapter but instead are derived from the quality of the data.

In fact, as explained in the first section of this book, the use of behavioral data to inform the modeling of ABMs of social phenomena can provide many relevant analytical advantages. The usage of the data has to be done with care and attention, but if the starting data is not "good enough," there is no technique of analysis or modeling that can fix that problem.

For instance, one of the most crucial problems under the approach of behavioral CSS is the one of the resulting model robustness. In fact, even a model with a high level of empirical salience and with much analytical power in explaining the causal mechanisms generating some social phenomena of interest can be of limited utility if used to study different phenomena and institutional changes.

A model, in other words, cannot be validated and be robust *per se* but only in terms of its use. From the opposite perspective, the problem can be an inappropriate use of the model, not its robustness and validation.

To make the point clear and to connect the discussion with the process of calibration and validation over behavioral data, we can make some examples. Imagine to start from an ABM validated over data regarding behavior and interaction and capable of supporting the explanation of the social phenomena for which it was designed. If, for instance, we modify the timing within the model, for instance, trying to use a model based on daily interactions to study a decadal dynamics, we are not sure that the model will still be appropriate. Similarly, it happens if we are studying policy proposals or institutional changes that could potentially modify behavior and interaction in unforeseen ways.

In all these cases, the model would probably lose its robustness and analytical power. Whenever there are drastic changes in analytical aims, the model validation has to be reassessed, and often, this requires the collection of new data.

The techniques presented in this chapter in fact are necessary for purposes of modeling and calibration, but they are only one element of validation activities that need to be done on ABMs. In fact, the calibration and classification activities presented earlier do not guarantee the validation of the entire ABM. However, once the behavioral models are ready, it is later possible to validate behavioral models in a way similar to what described presenting classification.

In fact, since the entire sample (i.e., all individuals) has been classified to different behavioral models (or behavioral models have been calibrated), it is possible to compute an aggregated measurement of error that quantitatively describes the capability of modeled behavior to describe the empirical data. It is also possible to use statistical tests to verify if there are significant differences between modeled and observed behavior.

Further, validation of behavior can be extended in order to consider interaction structures (see Chapter 7) and ultimately can be verified in terms of explanatory power of the resulting ABM and by means of sensitivity analysis.

The validation of a model, either in its behavioral components or in others, is thus much dependent on the data that is used. Further, the data and the model have to be consistent with the intended use of the model and with the focused empirical reality (Boero and Squazzoni 2005). If the model aims at understanding a heterogeneous population, not only the considered behavioral models have to able to capture the heterogeneity in the population, but also empirical observations have to support that (and data collection should be designed taking heterogeneity into account).

Finally, the techniques presented in this chapter allow developing models of behavior that have to be implemented later in ABMs. The phase of model implementation has to consider a few specific problematic issues. The first one regards discrete decision variables. Sometimes, the regression models that are used to study behavior have an intrinsic linear nature, while observed decision variables are different. Outputs of behavioral models in ABMs have to be transformed with care, without losing accuracy because of approximation. Similarly, with nonlinear models such as multinomial logit, the model predicts a cardinal value that has then to be evaluated over the cut points delimiting categorical values. Cut points are thresholds and have to be included in the modeling. If, for instance, the model predicts 2.4 and the threshold for choosing action "A" instead of "B" is 2.2, the modeling of behavior in ABMs does not only include estimated coefficients but also estimated cut points.

Furthermore and most importantly, when using regression models, it is worth implementing the entire regression model in the ABM, including coefficients and variables that are not significant but that improve the accuracy of the model. The possibility to include the standard errors of coefficients in ABMs is particularly interesting since it provides models with a probabilistic modeling of behavioral uncertainty that propagates through the entire model itself and that can be quantified and analyzed.

6

Sophisticated agents

When the analysis is conducted in order to explain the evolution of behavior in conjunction with the social phenomenon, more sophisticated models of agents are often needed. In fact, the explanation of the driving forces behind behavior can sometimes enrich the comprehension of social phenomena. When this is the case, it is possible to go beyond the behavioral modeling of agents and to provide agents with a modeling that allows investigating the causal forces that make behavior evolve.

The difference in analytical power between sophisticated agents and behavioral ones derives from the different sets of causal relationships that the two approaches allow modeling and investigating. In the case of behavioral agents, causality regards the decision-making process, that is to say the map that exploits available information to decide the course of action. In sophisticated agents, on the contrary, causality regards higher-level processes, mostly in the cognitive domain, that determine the map of causality defined in

Behavioral Computational Social Science, First Edition. Riccardo Boero.
© 2015 John Wiley & Sons, Ltd. Published 2015 by John Wiley & Sons, Ltd.

behavioral agents. In other words, the reason for choosing sophisticated agents is the interest in investigating why agents behave in a certain manner.

Examples of useful applications of sophisticated agents in behavioral CSS are those where social phenomena take long times to emerge and where the evolution of agents' behavior along social interactions are fundamental to fully comprehend dynamics and possible outcomes of exogenous changes, such as new policies and other shocks.

The tools we present in this chapter are exactly aimed at modeling more powerful and sophisticated agents in ABMs. In the ABMs community, these tools are often referred to as the "cognitive" approach to agents, while behavioral approaches are sometimes called the "reactive" approach (Sawyer 2005). Sophisticated models of agents are aimed at modeling the "mind" of agents, mimicking their cognitive functions and structures.

The identification of precise labels for defining sophisticated agents, however, is difficult because of many reasons. Most importantly, definitions in this field are always overlapping. For instance, it is undoubtful that all agents used in ABMs are somehow reactive. In fact, agents in ABMs always behave depending on some inputs received from the environment and in social interactions.

In order to avoid confusing definitions, we follow in this chapter the same principle adopted in the rest of the book, and we thus categorize and present tools according to the perspective of modeling and analysis. In particular, the rest of the chapter presents more sophisticated agents in an increasing order of complexity from the modeling perspective. Two main categories of sophisticated agents are presented. The first one includes tools that allow modeling cognitive processes without modeling the cognitive structures underlying them. In other words, they consider only the cognitive outcome, that is to say the process or the activity and not the mental structure. The second category considers also the cognitive structure.

Further, all the tools presented here are useful because they provide researchers with much significant aid in model development and in analytical capabilities. However, they are not the only available options. In fact, besides the fact that new tools are developed and emerge every day, the flexibility of ABMs allows modeling any kind of theoretical or empirical model in agents. Thanks to the flexibility provided by ABMs, it is thus possible to recur to *ad hoc* models of agents. The need or the preference for *ad hoc* models should be determined by analytical needs and by the availability of robust and established tools.

Before introducing the tools, we discuss their common features.

6.1   Common features of sophisticated agents

All the tools we present here follow a bounded rationality approach. In particular, all consider limited information about the world and imprecise and bounded information about others' actions and intentions. Most of the tools implicitly or explicitly assume also limited cognitive capabilities.

All the agents described here can also be defined as "adaptive" agents. In fact, the main purpose of using sophisticated models of agents is studying how they change and evolve behavior according to the stimuli coming from their social environment and in conjunction with their cognitive capabilities and eventual aims. Similarly to what has been said before about reaction, we prefer not using the label of adaptation to group all these approaches because of our focus on modeling and because adaptation can often be confused with optimization and learning.

From the modeling perspective, all the approaches replicating a cognitive process or/and eventually a cognitive structure share two common features. The first one regards the role played by the rest of the ABM (i.e., the part "outside" agents). The rest of the ABM defines the space in which agents live, and it thus provides all the information available to agents. The second feature regards the broadly defined interface between the models in agents and the rest of the ABM. In fact, agents and models of their behavior receive stimuli or inputs from the rest of the ABM. Those inputs are information about the environment and on what other agents are doing. More complex models can also include the perception of those information pieces. At the same time, the output of agents affects the rest of the ABM: in fact, agents decide and act in the environment. In other words, the sophisticated agents included in the model codetermine the environment that shapes their evolution. The use of sophisticated agents is thus justified exactly when the analytical need is to investigate that complex interacting dynamics.

Finally, most of the tools presented here have been developed by means of interdisciplinary efforts. The scientific domains that have most contributed to this field are artificial intelligence in general and machine learning in particular, evolutionary game theory, and social and cognitive psychology.

6.2   Cognitive processes

As mentioned previously, there are tools at disposal to model cognitive activities without modeling the underlying mental structure (that gives raise to the activities themselves). The fact of not considering the underlying structure and focusing on the process does not deny the existence of the structure. It just simplifies modeling.

In the following, we present three main modeling approaches. The first one focuses on reinforcement learning with simple processes. The second one considers other kinds of adaptation with bounded rationality. The last one introduces some nature-inspired algorithms that sometimes are used to model adaptation.

Two further points should be considered. First, from the analytical perspective, models are often not true alternatives. In fact, for instance, reinforcement learning can also be modeled with nature-inspired algorithms. As an example, Weisbuch et al. (2000) studied a model with a simple reinforcement learning that has been extended and made more sophisticated in Kirman and Vriend (2001) by using the nature-inspired algorithm called learning classifier system (LCS) . Further, reinforcement learning can also be modeled with cognitive structures presented in the following section and with *ad hoc* models.

Second, from the perspective of modeling, there are some cognitive processes that are particularly difficult to model with the approaches presented here because they do not consider the underlying mental structure. It is, for instance, the case of processes of social intelligence that require an internal model of the world, of the environment, or of other agents. With cognitive processes alone, this can be achieved only for simple environments where the internal representation of others is limited. Examples can often be found in evolutionary game theory and in studies of industrial organizations where the behavior of others can be reduced to a few strategies (e.g., Kirman 2010).

6.2.1   Reinforcement learning

Algorithms and tools to model reinforcement learning in agents should not be confused with tools aimed at optimization on empirical data. Those, like the ones presented in the preceding chapter, are aimed at calibrating the model of the agent on the observed behavioral pattern by means of optimization. These, on the contrary, are tools that allow modeling in agents the process of learning on incomplete information.

Further, these tools are not intended to study learning processes from a psychological perspective (though ABMs can be used to study developmental psychology; see, for instance, Abrahamson and Wilensky 2005) but to support the investigation of the role played by learning in social dynamics.

The learning process mimicked by these tools is unsupervised. Learning is achieved through a long path of experience and exploration of possible alternative behavior. Though unsupervised, learning requires feedbacks from the environment, which in this context is the rest of the ABM. Feedbacks in these tools take the form of environmental awards or punishments, following a behavioral approach (Boero and Novarese 2012). Feedbacks derived from reflections, conscious knowledge, and mental models require a cognitive structure, and thus, modeling learning upon that kind of feedbacks can be achieved with the tools presented in the following section.

Learning modeled with these tools is often referred to as individual learning unless agents' actions and received feedbacks are made public and learning becomes social. However, that definition can be equivocal since in this context the external environment (i.e., whatever is outside the agent) is a social environment, and thus, feedbacks are social products and learning is always social to a certain degree. Further, the same agent modifies the (social) environment with his own actions, and thus, the external environment is not independent and exogenous. Because of those complex dynamics of coevolution, the resulting environment is probably much noisy, and thus, reinforcement learning is often a good choice for both modeling and adaptation.

The tools to model cognitive processes of reinforcement learning are mainly the outcome of the close interaction between research in machine learning and evolutionary game theory. Following the discussion in Shoam et al. (2007), in this tradition, it is possible to identify three main modeling strategies: the modeling of others' strategies, the use of reinforcement learning without models of others, and the minimization of regret in playing sequences of strategies over given choices by others. The tools we present here mostly belong to the latter two approaches. Similarly, it is possible to identify four research agendas or aims that justify the usage of learning in this field (Shoam et al. 2007). They all partially overlap with reasons to include sophisticated agents in ABMs built according to the approach of behavioral CSS. The first research aim is computational, and it aims at understanding features of the social model that cannot be derived analytically such as solutions and equilibria in game theoretical dilemmas. The second one is descriptive, and it investigates which models of learning are most suitable to describe or explain empirical and experimental evidence. The third use is normative, and it aims at investigating the robustness of equilibrium solutions over longer periods and in cases of exogenous shocks. The last use is prescriptive and aimed at identifying learning processes that promote decentralized control structures. However, in most of the economic literature concerning game theoretical dilemmas, reinforcement learning is applied for the coordination toward a single equilibrium, that is to say as a selection mechanism (Kirman 2010).

Getting to the description of the tools, they mostly take the form of simple algorithms and they rely on the representation of two psychological laws, the law of effect and the power law of practice.

The first algorithm starts by enumerating all the possible course of actions for agents (i.e., the possible strategies or behavior). This obviously implies that all possibilities have to be identified ex ante by the modeler and that learning collapses in selecting the most appropriate behavior among a limited and predefined set of possibilities. The lack of capability to discover a new possible behavior is a further element contributing to the simplicity of these tools.

The second component of the algorithm is that for each possible behavior, there is attached a probability of being selected. Usually, at the beginning of the simulation of the ABM, such probabilities are equal for all possibilities.

The third component is the update of probabilities according to the feedback received. Such feedback and thus also the update of probabilities are usually done on each interaction and only for the single behavior that has been experienced. As mentioned previously, the feedback is an award or a punishment or in general a payoff that comes from the environment.

The fourth element of this first algorithm is then the normalization of probabilities. In summary, there is a probability number associated with all possible courses of action, and it represents the normalized (over the sum of all feedbacks received) sum of payoffs received when that course of action has been selected. The probability of selecting a specific course of action increases if that action has generated positive feedbacks relatively to other courses of action experienced in the past.

The process, as depicted previously, has infinite memory, and there is very much path dependence. In fact, if perturbations are not added to the model, such as including in the algorithm the need for trying all possible actions in the first rounds, the exploration of the solution space can be incomplete.

The algorithm replicates the law of effect because actions that in the past have generated a positive feedback are more probable to occur in the future. Also, the power law of practice is respected because the initial exploration of the solution space is more valued. In other words, feedbacks that have been received at the beginning modify probabilities by measures that are larger in relative terms than subsequent feedbacks, even if these latter are of the same absolute value as the first ones.

The second kind of reinforcement learning, similar to the first, is the one that minimizes the regret associated with choices by comparing possible actions over a set of recently experienced environmental conditions. Differences with the algorithm depicted previously are that memory in this case is often bounded to a limited set of recent history and that it requires agents to have a model of the world so that, first, they can separate others' behavior from theirs and from the environment in recent historical data. Second, it also requires agent to "simulate" the model of the world with different possible actions in order to minimize the regret, under the assumption that others would not react to changes (or that reactions of others are perfectly known). Because others' reactions often remain unknown or their knowledge is largely imprecise, learning is not straightforward and the algorithm is, again, reducible to a selection mechanism. An example of this modified algorithm is the seminal work studying the El Farol Bar problem in game theory (Arthur 1994), where the author's aim was explicitly to emulate human process of induction in strategy selection (the use of more sophisticated algorithms can provide further functionalities and processes even in this simple game theoretical setting—e.g., the use of genetic programming in Rand 2006).

However, the use of limited memory makes this algorithm not respect the power law of learning since rewards to actions are considered only over a limited window of time.

The third, a little more complicated, algorithm of reinforcement learning is the one introduced in Roth and Erev (1995) and Erev and Roth (1998). The agents employing these algorithms are sometimes called "Roth–Erev agents" or agents using the three-parameter model of Erev and Roth.

The algorithm is similar to the first introduced previously but for the introduction of two more parameters. The first added parameter is a forgetting value that decreases cumulative rewards when feedbacks are received. The memory loss equally impacts all possible courses of action, not only the one just explored. The second parameter that is added is one that allows sharing feedbacks between similar courses of action. The parameter is actually multiple and measures the degree of similarity between possible couples of actions. It is also bounded in the interval between 0 and 1. If it equals 0, it means that the two actions are completely different, and if it is equal to 1, it means that the two actions are identical.

The consequence of these modifications is that path dependence in learning is less pronounced since probabilities of actions not directly experienced can still be modified and actions will not be excluded prematurely. However, the cost to accomplish that is the addition to the model of an *ad hoc* structure of similarity between courses of actions that is both difficult to be specified and validated by researchers and that is often implausible to assume taking into account limited knowledge and cognitive capabilities at the individual level.

A fourth and last kind of reinforcement learning can be obtained by considering social learning. In fact, all previous algorithms can easily be modified by making public the selection of actions and the reception of feedbacks (e.g., Hedström 1998). The publicity of that information can be universal or related to social interactions, so that it is shared among social neighbors. When social learning is adopted, reinforcement learning also replicates processes of imitation, and in fact, such an approach has been used in studies of innovation emergence and diffusion, where learning concerns the exploration of the technological landscape, the identification of successful products and business strategies, and their diffusion in the industry.

Main models of reinforcement learning are summed up in Table 6.1.

6.2.2   Other models of bounded rationality

Further modifications of simple algorithms for reinforcement learning can approximate other cognitive processes related to models of bounded rationality.

For instance, modifying feedback evaluation (e.g., Nebel 2011) toward the satisficing approach (Simon 1956) is relatively easy, as it is considering aspiration levels (Selten 1998) and their dynamics (Boero et al. 2004). In both cases, the adaptation process of agents is modified in particular from the perspective of the trade-off between exploration and exploitation of the space defined by possibilities of behavior. In fact, probability update rules need in these cases to accommodate the achievement, respectively, of the satisficing threshold and of the aspiration level.

Similarly, it is possible to study search, stop, and selection rules under the hypothesis of the adaptive toolbox and of ecological rationality (Gigerenzer and Selten 2001, Smith 2003). For instance, the entire adaptive toolbox can be represented in ABMs (Wittmann 2008), or reinforcement learning can be used to model information selection (Todd and Dieckmann 2005) and heuristics selection (Boero et al. 2008), even taking into account the approach of cognitive niches (Marewski and Schooler 2011) or the learning theory of strategy selection (Rieskamp and Otto 2006).

6.2.3   Nature-inspired algorithms

The algorithms described previously have a very simple structure from both the modeling and computational perspective. In fact, they are often *ad hoc* modeled in ABMs.

More sophisticated models of adaptation require more modeling and computational effort. In particular, even if external tools are used to facilitate modeling, the development of ABMs with agents behaving according to them is difficult because of the more complicated interface with the rest of the

Table 6.1 Main simple models of reinforcement learning and their characteristics.

Model	Working mechanism	Main features
(1) Selection on experience	(a) Update probability of selected choice with received feedback (b) Normalize probabilities (c) Select new choice on probabilities	1. Law of effect 2. Power law of practice 3. Infinite memory 4. Limited exploration of solution space
(2) Regret minimization	(a) Record choices and feedbacks in recent past (b) Simulate the feedbacks that would have been received with different choices (c) Select the choice that would have received the best feedbacks	1. Law of effect 2. Limited memory 3. Wide exploration of solution space, based on model of the world
(3) Roth–Erev agents	(a) Decrease all cumulative rewards (b) Update probability of selected choice with received feedback (c) Eventually, update probability for similar choices (d) Select new choice on probabilities	1. Law of effect 2. Power law of practice 3. Limited memory 4. Wide exploration of solution space, based on model of similarity between choices
(4) Social learning	As in (1) or (3) but information about selected actions and feedbacks is public	1. Law of effect 2. Power law of practice 3. Infinite or limited memory 4. Wide exploration of solution space, based on social sharing of choices and feedbacks

ABM. Although some design solutions have been proposed in the literature (e.g., the "Environment–Rules–Agents (ERA)" scheme in Terna (2000) and Gilbert and Terna (2000)), we focus here on the introduction of the basic working mechanisms of these algorithms in order to provide the reader with the intuition of the several analytical capabilities they provide.

The algorithms we refer here are all inspired by nature in a broad sense. In fact, they mimic processes of natural evolution, insects and other animal societies, or the structure of physiological organs such as the brain.

Nature-inspired algorithms are heuristics for optimization particularly useful when the solution space shows features that make standard optimization fail. They can be used to replicate several cognitive processes but still relying on the capability of these algorithms to replicate reinforcement learning. Nature-inspired algorithms usually do not require the modeler to define all possible solutions ex ante. Nevertheless, they require a broader but bounded and precise definition of the solution space.

The first family of nature-inspired algorithms worth mentioning is the one that mimics processes of evolutionary genetics. Algorithms exploit a parallel and multiple exploration of the solution space similar to what is achieved with social learning. Moreover, they adopt mechanisms to find new possible solutions to explore. Among those algorithms, the most important one is called genetic algorithms (GA) (Holland 1975).

At the basis of GA, there is the modeling of a single solution candidate. It is achieved by using a list of characteristics, usually represented as a string made by bits taking value 0 or 1. Each position in the string is like a gene in DNA, and it represents some features of the agents' behavior that can be active or not (usually 1 represents activation, 0 otherwise). Strings, which are possible solutions, can be interpreted by a function, called the "fitness" function, that computes a feedback. In other words, the fitness function computes the environmental reward for the behavior modeled by the string. The presence of a fitness function in agents implies the cognitive capability of modeling the environment.

If the GA ended at this point, the algorithm would collapse to an algorithm for heuristic selection very similar to the ones presented before. GA, on the contrary, mimics also the evolution of genetics, and thus, solutions live only a generation, and after that, another generation is created to better explore and evaluate the solution space.

In fact, solutions are reproduced in the following generation according to a set of mechanisms that rely on the feedback obtained from the fitness function.

The first mechanism for reproduction is just a random wheel that gives a different probability of reproduction to solutions depending on their fitness. The fitness values thus means fitness to reproduce or survive.

A second mechanism randomly selects better solutions (i.e., those with higher levels of fitness) and then applies mutations to their genetic code. This means to randomly switch some of the components of the strings so that those that were 0's in parents become 1's in children and vice versa.

The third mechanism, usually called "crossover," randomly selects couples of solutions according to their fitness (i.e., by means of a random wheel that gives more probability of selection to solutions with higher levels of fitness), and then, it creates two child solutions mixing the strings of the two parents. The crossover of "genetic" material usually happens by randomly selecting a cut-point in the string (i.e., a position, a gene, in the string). The two parent strings are thus separated in two substrings each cut at that point. The first child is made by the first substring of the first parent and by the second substring of the second parent. The other child solution is made by the opposite, which is the first substring of the second parent and the second substring of the first parent.

Mutation and crossover of solutions allow exploring solutions unknown and unexplored. The reproduction of successful solutions as they are (i.e., without mutations and crossover) allows memory and accumulation of learning.

Even if using GA in agents implies a model of the world and it partially resembles social learning as discussed previously, still, GA can be extended to explicitly allow social learning by sharing information about the exploration of the solution space. In other words, when using GA, it is still possible to extend learning to the social level by allowing agents to exchange information about the best solutions they have found. Similarly to what has been discussed before, individual and social learning with GA can lead to completely different outcomes (Vriend 2000). Again, with a high degree of similarity to what has been said when discussing simple reinforcement learning, GA has been extensively used in studying innovation processes (e.g., Engler and Kusiak 2011).

Another algorithm of the same family is called LCS (Holland et al. 1986). The main difference with GA is in the representation of solutions: they are sets of rules that often allow a more intuitive representation of behavior. The rest of LCS is similar to GA, even if several variants of the LCS algorithm have been developed and some of them diverge from GA in particular for the computation of the probability of reproduction.

Both GA and LCS have been extensively used in economic modeling to represent learning (Arthur 1990, Holland and Miller 1991).

Another algorithm belonging to this family is genetic programming, which allows a behavioral learning similar to GA but a looser and less biased definition of the solution space of behavior by the modeler (Boero 2007) and algorithms of swarm intelligence such as ant colony optimization where the solution space is represented as a graph.

The last nature-inspired algorithm of particular interest for behavioral CSS is artificial neural networks (ANNs), which are connectionist models inspired by the central nervous systems. They are particularly useful when there is a large amount of data available for the approximation of an unknown behavior, and there is not much interest in understanding which cognitive processes and structures underlie such behavior.

In fact, ANNs can be effective in classifying and learning large amounts of data, but it is very difficult to understand why they are making specific decisions (Rand 2006). The modeler who intends to use ANNs needs to know that they require an interface with the rest of the ABM that considers the information coming from the ABM as the ANN input neurons and the possible actions that agents can take in the ABM as the ANN output neurons. The modeler also has to choose several parameters governing the learning process happening in the hidden layers of neurons of the ANN.

From a modeling perspective, ANNs are thus limited by the fact that the modeling of a large space of solutions (i.e., of possible behavior) can be impossible and that it has to be fully specified ex ante. ANNs can allow modeling learning and complex behaviors, but they remain obscure black boxes and they thus have a lower analytical power. Their use in behavioral CSS is suggested for cases when the causality of the limited set of observed behavioral patterns is well known, but it is unknown how agents shift between such patterns.

As a final note, nature-inspired algorithms can be mixed in ABMs, and thus, it is possible, for instance, to have agents with ANNs evolved by a GA. The mixed use of nature-inspired algorithms is often justified analytically by the need to overcome the limits of each single algorithm.

6.3   Cognitive structures

When there is need for the modeling of high-order mental processes, more sophisticated tools are at disposal. Their main characteristic, from the modeling perspective, is that they include a cognitive structure that works to provide several functions. The use of these tools is thus suggested when it is important to represent in ABMs several cognitive functions at the same time and when their interaction plays a crucial role in explaining behavior and the social phenomenon on focus.

Adopting the modeling of cognitive structures often carries on further problems in modeling social intelligence and in calibration. Similarly to what has been presented previously, such problems vary very much depending on the tool adopted, and they are introduced later throughout their presentation. In the last section of the chapter, we sum up the critical points that usually emerge in using sophisticated agents.

Some of the tools of cognitive structures follow a symbolic approach that allows modeling the manipulation of symbols; other tools do not allow that. A few tools are based on the connectionist approach to cognition; others consider higher-order cognitive structures. Some tools rely on folk psychology, while others have been validated at the neural level. In summary, the tools presented are very much different and their use is even more nonstandard. However, they can be categorized in two macro sets. The first one includes all the tools that aim at modeling a cognitive structure capable to provide a rather limited amount of cognitive functions. The cognitive structure they include is thus a subset of the whole, and it is identified in order to provide desired functionalities. The second set includes the few tools that are all purpose, broad, and comprehensive. Because of these differences, the first kind of tools can be called middle-level structures. The second kind of tools is usually referred to as "heavy" or "rich" cognitive models because of their comprehensive aim and high level of sophistication.

6.3.1   Middle-level structures

Among tools for middle-level structures, most of them adopt a folk psychology approach. These tools have mainly been developed in the fields of computer science and artificial intelligence, and the most known of them is the belief, desire, and intention (BDI) model (Bratman 1987).

BDI agents have multiple goals (i.e., desires) they try to accomplish by means of plans of actions (i.e., intentions). They also have knowledge of the world that can be imprecise or even wrong (i.e., beliefs). BDI agents are usually implemented within a framework that is called procedural reasoning system and that is constituted by a database and an interpreter. The database contains beliefs, desires, and "knowledge areas" that are the possible plans of action. The interpreter is an algorithm aimed at managing the system and the necessary actions, either internal (e.g., update of beliefs) or external (e.g., acting upon decisions). The main component of the interpreter is the inference mechanism that decides which action to undertake considering BDI.

The basic architecture of BDI agents does not explicitly allow learning and social interactions; however, the scheme is enough general to be extended in almost any direction. For instance, it is possible to extend BDI agents to the domain of emotions, introducing simple emotion models and allowing social agents to behave not only on the basis of internal goals but also on emotive reactions to perceptions (Parunak et al. 2006).

Further, BDI agents can look rather static and predetermined by the modeler, but actually, the model can be integrated with other tools such as GAs to obtain evolving and learning agents (for an application to financial markets with ABMs, see Costa Pereira et al. 2009).

Many software for the development of BDI agents are available today, either proprietary (e.g., JACK) or not (e.g., Jason); however, their integration with an ABM is always a demanding task because BDI software has been developed with the modeling of software agents and multiagent systems in mind and not for scientific agent-based models (Padgham et al. 2011).

In general, simple *ad hoc* models of cognition similar to BDI ones, either scientifically enrooted or based on folk psychology, can easily be integrated with other tools such as GAs and ANNs to model learning (Junges and Klügl 2013) and other phenomena of adaptation such as cultural change (Abrams 2013). Further, the use of inference mechanisms to manipulate rules representing BDI can be used for any kind of model of contextualized reasoning with ABMs (e.g., endorsements (Alam et al. 2010) or consumers' decisions (Jager and Janssen 2012)).

The last kind of tools for middle-level structures is the one of the social–cognitive approach (Conte 2002, Conte and Castelfranchi 1995, Conte and Paolucci 2001). It is an approach explicitly addressed at investigating the intersection between the social and the cognitive, and that has developed some frameworks aimed at investigating these issues with ABMs. Examples of frameworks and applications to social–cognitive issues are reputation dynamics (Conte and Paolucci 2002, Sabater et al. 2006) and social norms formation (Conte et al. 2013).

6.3.2   Rich cognitive models

Rich or "heavyweight" cognitive models for agents are only a few. More than tools, they are frameworks or architectures that define both a basic and irreducible cognitive structure and what cognition is. Structures, in turn, are implemented in several software tools, programming languages, and so on.

They are more complicated than the tools introduced before, they mainly derive from research in artificial intelligence, and they include different approaches, from symbolic manipulation to connectionist approaches, but they all share a few main characteristics that define the reasons to adopt them.

First, these tools allow considering goals, and thus, both cognitive processes and behavior can be goal directed. Second, they allow modeling physiological urgencies and emotions. Decision-making and perception of the world can thus be driven by the interaction of goals, physiological urgencies, and emotions, and thus, reflective and emotive cognitive processes can coexist and be investigated in conjunction (Kennedy and Bugajska 2010). Third, they allow explicitly considering cultural issues and social norms (i.e., differences at the macro level) as well as personality traits (i.e., differences at the individual level). In other words, they are more suited to model social intelligence since cognition models can capture both individual and social level processes and their interaction (Ron 2006).

From the modeling perspective, the main challenge they present concerns their heavyweight in terms of computational demands. In fact, similarly to other sophisticated agents, they require an effort to define and implement an interface between cognitive agents and the rest of the ABM. Further, in comparison with the other models presented previously, the cognitive structure that is modeled with these frameworks is particularly demanding in terms both of computer memory, either because of the size of the structure or because of the number of parameters and values representing internal states, and of computational power, because of the procedures to update internal states and to simulate decision-making.

Being frameworks more than simple tools, researchers who intend to use them need first to understand the frameworks' main characteristics and to study the language adopted (mostly LISP, but there are other implementations that are easier to be integrated with ABMs such as the ones of ACT-R in Java and Python). Then, researchers can build models within the framework aimed at implementing the particular instance of the cognition model that is needed. This means also that most of these models are sharable and reusable by others.

The first cognitive architecture that can be used is State, Operator, And Result (SOAR), and it relies on symbolic processing though most recent extensions allow also nonsymbolic processes such as reinforcement learning. The second architecture is Connectionist Learning with Adaptive Rule Induction Online (CLARION) that has a more connectionist "flavor" and it is based on many subsystems (e.g., action-centered subsystem, motivational subsystem) where knowledge is separated between explicit and implicit, allowing to explicitly address issues related to interactions within the dual representational structure. The last well-known cognitive architecture is Adaptive Control of Thought—Rational (ACT-R). It belongs to the symbolic systems approach, but it has been developed with much attention to the empirical results collected in the field of cognitive neuroscience. ACT-R allows considering procedural and declarative knowledge, and most recent versions of the architecture are divided in modules that can be mapped to different areas of the brain.

6.4   Critical issues: Calibration, validation, robustness, social interface

The use of sophisticated agents in ABMs means adding a complex layer of modeling to the ABM. When using the behavioral approach described in the preceding chapter, the agent model is simpler, and its specification is assumed to be constant and predetermined by forces such as learning, ecological rationality, or social norms. On the contrary, sophisticated models of agents allow investigating these particular processes of behavioral change by modeling cognitive functions and processes.

When the analytical need is to replicate behavior, cognitive models are not needed (Conte and Paolucci 2014). When the explanation of behavior emergence can enrich social explanation, the adoption of more sophisticated agents can provide the analytical power that is needed. The adoption of sophisticated agents, however, implies the need to face some methodological and analytical challenges.

A first challenge concerns the investigation of the model and the interpretation of results, and it is also related to calibration and validation. In fact, as a new layer of modeling is added to the ABM, a new set of possible dynamics and causes is also added, and understanding the model in terms of precise social mechanisms that take into account the several interactions between the model components can be more difficult.

At the same time, because of the relationship between sophisticated models within agents and the social dynamics of agents' interaction, it is often difficult to conduct sensitivity analysis and other analytical tasks on the model. This problem affects both the understanding of the model and its validation.

Further, the same problem affects calibration. In fact, models of agent cognition have first to be validated by themselves and then, when used in ABMs, to be calibrated. To make the point clearer, all the models introduced in this chapter require the specification of some parameters, and their number increases with their sophistication, going from a few (e.g., 1, 3 parameters) in simple reinforcement learning to tens or hundreds in rich cognitive architectures.

Moreover, while models for cognitive processes that ignore a cognitive structure seem attracting because parsimonious, they lack a referral to an empirical target from which to infer parameters for calibration. For instance, the several parameters that govern a GA cannot be inferred by any empirical data source, and the modeler is obliged to search in the parameter space for values that generate an adaptation process similar to what has been empirically observed.

In summary, models of cognitive functions are particularly difficult to calibrate because of the difficulty to collect data about cognitive processes. For the same reasons, it is difficult to validate them, and the lack of confidence on their structural validity does not allow evaluating their robustness, which is their reliability under different conditions and exogenous shocks.

For example, we know that reinforcement learning seems to explain several dynamics observed in long-lasting economic experiments (Erev and Roth 1998), but we do not know how to compare different models of reinforcement learning because we do not know models robustness (i.e., the limits of applicability of different reinforcement learning models).

Models of cognitive structures, on the contrary, and either if limited (i.e., middle-level structures) or if comprehensive (i.e., rich cognitive structures), seem to promise more robustness because it is theoretically possible to evaluate their structural validity (e.g., ACT-R can be compared with behavioral and fMRI observations (Anderson et al. 2008, Qin et al. 2007)).

However, today, there are not cognitive models that are fully validated and accepted as good models of cognition, yet. Further, even if a more limited aim is searched for, for instance, by considering only some specific subsystems of cognition, validation is still problematic due to problems of observation of cognitive processes mentioned before. In summary, even with rich cognitive models, calibration and validation are problematic and not only because of problems with the data but also because calibrating and validating those models can be excessively demanding for researchers in behavioral CSS.

In general, the use of sophisticated agents in ABMs often carries on a high degree of *ad hoc* choices that need much care. Such choices regard modeling, calibration, and validation. They impact on robustness and external validity of the model and of its results, and they are mainly driven by exogenous constraints such as the availability of computational power to run simulations and analyze results and the degree of acceptance in different disciplines of specific cognitive models.

To add a further critical element to this picture, the introduction of sophisticated agents carries on a further methodological problem at the point of the interface between the model in agents and the social interaction (Zoethout and Jager 2009). Except a few cases in the social–cognitive approach, cognitive models in fact have not been developed with the aim of studying social processes and social phenomena but to understand cognition processes within an individual.

Consequently, besides the difficulty in calibrating and validating cognitive models alone, there is also the difficulty to effectively execute the same tasks on the social interface between agents and the rest of the ABM. The interface in fact defines information that can pass between individuals, their possible actions, the emergence of social constructs (i.e., social norms), and the time-scales of information passing, action execution, and so on.

In order to address the analytical interests of behavioral CSS, it is needed to let cognitive agents and social interaction influence each other (e.g., a cognitive process influencing the structure of social interaction) through some mechanisms formalized in the social interface. The formalization of the social interface is particularly problematic, and it most likely results on a large amount of researchers' *ad hoc* hypotheses since the difficulty in collecting empirical data on those mechanisms and in relating the data to the adopted cognitive model.

7

Social networks and other interaction structures

Because social phenomena are social, they cannot be studied and explained by reduction to the individual level only. On the contrary, social phenomena require to explicitly take into account the fact that individuals are not isolated, that is to say to consider the essential role of social interaction.

Social interactions are many, pervasive, overlapped, and dynamic. For the sake of the analysis, however, they can be categorized in two main kinds. First, there are direct interactions in which social agents get personally in touch, although their interaction can be mediated by different means (e.g., language, technology, media, etc.). Direct social interactions usually allow participants to know the identity of counterparts, and thus, dynamics such as trust and reciprocation can easily emerge.

Second, there are indirect interactions by which we intend here interactions that happen through modifications of the environment in which social agents live. In this latter case of social interactions, participants cannot directly

Behavioral Computational Social Science, First Edition. Riccardo Boero.
© 2015 John Wiley & Sons, Ltd. Published 2015 by John Wiley & Sons, Ltd.

identify others, even if they very well know that modifications happening in the environment are due to the actions of other individuals.

A couple of examples make the distinction clear. Direct interaction can happen, for instance, between members of a family arguing or between friends participating to the same event. Similarly, a direct interaction happens in dynamics such as word-of-mouth diffusion processes where, for instance, the review of a new book is posted by somebody on social media. Readers of the review, although not knowing the reviewer personally, know who has written it and can further diffuse the review to other people, making the opinion their own or just referencing it.

From the opposite perspective, an indirect interaction is, for instance, the one that happens in financial markets. In fact, the market environment faced by agents is a social product determined by the actions of other agents (e.g., other investors), and participants to the market codetermine its dynamics every time they operate in it. The interaction structure of such markets is often regulated by complicated mechanisms, and even if agents have access to single bids and offers, such as in an order book, interactions are most often anonymous.

In the relatively short history of modeling in CSS, both direct and indirect social interactions have been considered. For instance, indirect interaction is at the basis of the pioneering ABM called Sugarscape (Epstein and Axtell 1996). Similarly, indirect interaction through modifications of space is at the core of modeling tools such as Cellular Automata that represent, from some viewpoints, precursors of ABMs.

From the opposite perspective, direct interaction through social networks is today one of the most common features in social simulation. Further, today, both social network analysis (from now on, SNA) and the so-called science of networks are approaches often considered belonging to CSS. In particular, SNA is a set of descriptive statistics that allows studying characteristics of social networks, and research in the science of networks exploits tools derived from SNA and from physics in order to establish relationships between social phenomena and the social network structure.

From the perspective of behavioral CSS and thus for the focus we adopt here, it is essential to consider two important issues related to SNA. First, in behavioral CSS, statistical tools of SNA are fundamental for the understanding of the essential characteristics of social networks allowing researchers investigating realistic social interactions. Second, it is essential to be able to model such characteristics in ABMs through opportune and reliable algorithms.

Because of these needs of behavioral CSS, the first two sections of this chapter initially focus on basic SNA statistics and then on the algorithms to generate in ABMs social networks with the specific characteristics that have been observed empirically.

Then, the chapter proceeds with a section that shortly discusses other kinds of interaction structures, either direct or indirect ones. In comparison with social networks, other interaction structures are most of the time *ad hoc* structures that require a specific analytic and modeling effort. That section discusses the most used approaches to accomplish that.

The final section of the chapter discusses the two most important and critical points related to any kind of interaction structure modeled in behavioral CSS, that is to say the temporal dimension (i.e., dynamics of social structures) and the relationship with behavior.

7.1   Essential elements of SNA

The first requirement to study social networks is to represent them formally through matrixes and graphs. Matrixes, often referred to in this context as "sociomatrixes," are used to manage large amounts of data and to compute statistics on it. Ultimately, sociomatrixes are useful for analyzing the characteristics of empirically observed social networks. Graphs, on the contrary, are used to visually represent social networks, and they allow summarizing some of their characteristics. Further, the representation through graphs is usually the one most adopted in modeling (see the following section).

We consider a simple example of social network made by six agents, labeled with capital letters from A to F. Its sociomatrix is presented in Table 7.1. Rows and columns of the matrix represent agents, that is to say individuals or organizations. In the cells of the matrix, the value 1 is presented if there is a link between the agents referred by the row and the column where the cell is. The value 0 is used otherwise. Cells along the diagonal of the matrix are left empty

Table 7.1   An example of sociomatrix.

	A	B	C	D	E	F
A		1	0	0	0	0
B	1		1	1	1	0
C	0	1		0	0	0
D	0	1	0		1	0
E	0	1	0	1		1
F	0	0	0	0	1	

because in this case it does not make sense considering a link that originates from an agent and directly arrives at the same agent.

There can be different structures of sociomatrixes. The one presented here uses agents both in columns and rows, but there are other possibilities. For instance, if the focus is on considering the network resulting from the participation to a series of events, agents can be represented along one dimension and events along the other. In this latter case, a cell with a 1 represents if the agent has participated to the event and 0 otherwise.

Moreover, it is easy to notice that the sociomatrix in Table 7.1 is symmetrical. The reason for that is that we are considering here an example of social network with undirected relations between agents. That is to say that the relation between a couple of agents, if present, is symmetrical. Otherwise, we would have a social network where, for instance, agent A can influence agent B (i.e., a 1 in cell A, B), but not vice versa (i.e., a 0 in cell B, A). Further, it is possible to attach a weight to the direct relationship between two agents by using multiple values of 1 or continuous nonnegative values.

The example of social network formalized in Table 7.1 is represented as a graph in Figure 7.1. As said earlier, we are considering here an undirected social network, and thus, we present an undirected graph. When directed relations are present, they are usually represented in graphs as arrows and not just as line segments in order to provide information on the direction of the link between agents. Similarly, when relations are weighted, it is common

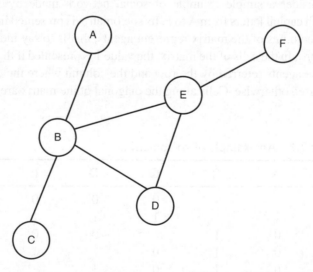

Figure 7.1   The example of undirected social network of Table 7.1 presented as a graph.

practice to graphically represent the weight of a link with its thickness (e.g., links with bigger weight are presented thicker).

SNA provides social scientists with so many measurements that it is almost impossible to compile a comprehensive list. However, the ones most used can be classified in three macro categories (for a complete introduction to the formalization of social networks and to their main statistical measures, refer to Wasserman and Faust (1994)).

First, there are measurements or statistics related to single nodes, usually called "node statistics." Second, there are measurements of characteristics of the network taken as a whole, called "network statistics." Last, there are statistics referring to groups of nodes smaller than the entire network, and they are labeled "group statistics."

Among node statistics, there are measurements of how much a node is connected to others or, in SNA jargon, of its centrality. The degree of a node is the most used measurement of this kind, and it is exactly the number of other nodes to which a node is directly connected. It is a measure of centrality because it tells how many other nodes can immediately be reached by a node.

For example, in our network, the degree of node B is 4, as can be seen in Figure 7.1 and computed in Table 7.1 by summing the values in all the cells of the column (or the row) labeled "B."

Further, degree measurements can be extended. For instance, the centrality of node B can be evaluated by considering also the nodes that can be reached indirectly in two steps, as, for instance, node F (that is reachable through node E in two steps). The second degree of node B in our example is thus 5 because in that network all other nodes can be reached by node B in two steps. It is much different for nodes A and C that have a degree equal to 1 and a second degree equal to 4.

In comparison with B, nodes A and C look more isolated but not too much thanks to the second degree. However, the example points out also the difficulties associated with the comparison of descriptive statistics. In fact, node statistics do not consider the size of the network. The evaluation of node statistics, thus, is often troublesome (i.e., is that value high or low?), and it requires a number of comparisons that increases nonlinearly with the size of the network.

Further, as in any other field, statistical results have to be interpreted, but interpretation here is particularly difficult. For instance, in our example, we know that nodes A and C could be not so "peripheral" if we consider the link they have with node B and the second degree. Unfortunately, however, we do not know whether node B is prone to be used as an intermediary by A and C. To say in other words, whenever structural measurements are not only described but they need to be interpreted, the behavior of nodes (e.g., what node B will do) defines the interpretation of measurements. Without behavioral information, structural properties have no analytical meaning. In our example, a "hostile" node B allows interpreting degree and second-degree measures of nodes A and C as indicators of noncentrality in our small network. On the other hand, a "cooperative" node B allows interpreting the same measures in the exactly opposite way.

The example just discussed strengthens the need to recur to a modeling approach of social phenomena that includes both social interaction and the behavior of agents.

Furthermore, it is often useful to study the distribution of degrees, which is the description of the probability of a node to have a specific value of degree, and the average degree, which is the expected degree value for nodes in the network.

Besides nodal degree, the distance between nodes is another measure worth considering. The distance between two nodes is often called the geodesic distance, and it equals the path made by the minimum number of links that are required for going from a node to another one. Knowing distances between nodes allows evaluating how much it takes to reach nodes, or their reachability.

In our example, the distance can be easily observed in the graph presented in Figure 7.1, but it can also be computed using the sociomatrix presented in Table 7.1. The computational approach using matrixes is particularly useful when networks are large and connection paths are multiple.

The computation of nodes' reachability is based on the products of the sociomatrix, that is to say its powers. When raising the sociomatrix to power 2 (Table 7.2), we obtain a new matrix where all nonzero cells mean that there is a path of length 2 connecting the nodes (i.e., the couple of nodes represented by the row and the column of the cell).

Table 7.2   Sociomatrix of Table 7.1 raised to power 2.

	A	B	C	D	E	F
A	1	0	1	1	1	0
B	0	4	0	1	1	1
C	1	0	1	1	1	0
D	1	1	1	2	1	1
E	1	1	1	1	3	0
F	0	1	0	1	0	1

Table 7.3   Sociomatrix of Table 7.1 raised to power 3.

	A	B	C	D	E	F
A	0	4	0	1	1	1
B	4	2	4	5	6	1
C	0	4	0	1	1	1
D	1	5	1	2	4	1
E	1	6	1	4	2	3
F	1	1	1	1	3	0

To make the point clearer, we know from Table 7.1 that node A is not directly connected to node E; in fact, in that table, the corresponding cell has value 0. In Table 7.2, we note that there is a path of length 2 connecting A with E and vice versa; in fact, the corresponding cell value is 1.

Similarly, we can raise the sociomatrix of Table 7.1 to power 3, and we obtain the matrix presented in Table 7.3 where all nonzero cell values represent the presence of a path of length 3 connecting the corresponding nodes.

For instance, nodes A and F that were not connected with direct links (Table 7.1) and with paths of length 2 (Table 7.2) are now connected with a path of length 3 (Table 7.3).

By computing the powers of the sociomatrix, we can thus discover the minimum path required to connect any couple of nodes in a social network.

Moreover, raising a sociomatrix to its powers provides a further important information. We have said that the existence of a link of length equal to the power is represented by a nonzero value in the corresponding cell. The precise (nonzero) value in that cell tells us the number of possible paths that exist in the network with length equal to the power of the matrix.

Consider, for instance, the value 4 in the cell in Table 7.3 related to nodes A and B. It means that there are four possible paths of length 3 connecting node A with node B and vice versa. Looking at Figure 7.1, we know that starting from node A, we have to go, as the only possible first step, to node B. Then, as a second step, we can go back to node A or go to nodes C, D, and E. Finally, from the four possible nodes reached at the second step (i.e., A, C, D, and E), we can reach node B with a third step. As this example testifies, there are actually four possible three-step paths for going from node A to node B, confirming what presented in Table 7.3.

With statistics such as nodal degree and shortest paths, it is possible to compute almost all other SNA measures even at the network and group level.

For instance, as a network statistics, it is possible to compute the average nodal degree in a network and compare it to the maximum level possible, which is $N-1$ with N equal to the number of nodes in the network.

We can also compute the diameter of the network, which is computed as the "maximum" shortest distance present in a network. In fact, the diameter of a network is evaluated by first computing the shortest distances between any couple of nodes in the network, as described earlier. Second, the maximum value of these distances is selected. If the network is not fully connected, which is the case when subnetworks can be separated without breaking any link, the diameter is infinite (West 2000).

Finally, because we are particularly interested in social networks with a high level of empirical salience, we consider the statistics called clustering coefficient. The clustering coefficient can be global, at the network level, and in this case, it is sometimes called "transitivity" (Wasserman and Faust 1994). It can also be a local clustering coefficient that measures clustering at the node level. Between the two definitions, the latter is preferable because it is more useful when the description of the network is incomplete and because it still can be used to compute measures of the characteristics of the network as a whole by using the average local clustering coefficient.

The local clustering coefficient (Watts and Strogatz 1998) is the number of links existing in a group divided by the maximum possible number of links that could exist in the same group. Groups are, in this context, usually referred to as neighborhoods, and in an undirected graph, the maximum number of possible links in the neighborhood of node i is $(k_i(k_i-1))/2$ where k_i equals the number of neighbor nodes.

The neighborhood of a node is usually identified as the set of nodes directly connected to that node. The local clustering coefficient thus measures how much the neighbors of a node tend to be fully connected. A high local clustering coefficient is a very common characteristic of empirical social networks.

In conclusion, there are several specific software tools available to conduct SNA and for visualizing networks (e.g., UCINET, Pajek, NetDraw) and several packages and libraries for common-purpose statistical packages (e.g., R and Stata) and programming languages (e.g., NetworkX for Python). However, the matrix representation and the statistics described earlier do not require any particular or advanced software.

As mentioned before, the statistics we have introduced in this section allow capturing the most essential structural properties of social networks. At the same time, the graphic and matrix representations allow formally managing social networks and studying them. However, there are still some remarks that need to be made regarding how data about social networks can be observed and measured.

Data collection about social networks is strongly supported by the large availability of data collected by computers according to the means and approach usually referred to as "Big Data." However, there are social networks for which scholars have to design specific *ad hoc* surveys that require much care.

In fact, besides the usual care required in survey design, social networks pose unusual challenges related to the representativeness of the sample whenever surveying the entire network is not feasible because of excessive costs or of any other reason.

Developing representative samples for surveying social networks presents two problematic requirements. First, the relationship between nodes' features in the social network (e.g., their reachability, centrality, etc.) and their sociodemographic characteristics should be known, so that the latter, which are usually the only ones at disposal ex ante, can be used to draw a sample. Second, knowledge about the structural properties of the universe network is needed both to draw the sample and to evaluate the uncertainty and error associated with the measurements on the sample. Knowledge about both those features is most often not available and its collection unfeasible.

In order to solve this kind of problems, a first solution is to consider the representativeness of the used sample with care: it can be representative of other dimensions such as the sociodemographic ones, but the representativeness of network structure remains to be verified by means of replication and extension of the analysis. Second, some different sampling techniques specifically designed for social networks can be employed, such as "snowball" sampling described in Atkinson and Flint (2001) or the ones reported in Erickson and Nosanchuk (1983).

7.2   Models for the generation of social networks

When modeling social networks in ABMs, there are two possibilities depending on the kind of empirical evidence at disposal.

The first option is available when the entire network has been observed, and thus, the dataset at disposal allows replicating the social network entirely and with the highest level of detail. It is, in other words, the case when all nodes and all links between them are known. In this case, all ABM platforms make available specific libraries, objects, or functions to replicate the network structure as it has been observed. In particular, tools allow transforming agents into nodes of the network. Links between nodes/agents, sometimes called edges, can be created as well, either directed or undirected and with any weight or other characteristic.

The second option is required when only some information about the network is at disposal. Usually, such limited knowledge of the social network is made by the average nodal degree, the shape of the distribution of degrees, the network diameter, and clustering coefficients.

When those characteristics are the only ones known, it is impossible to replicate a precise social network, and thus, it is required to randomly create a social network with similar characteristics. To say it differently, the need is for a network with features similar to the ones observed in reality, and that is the same as randomly selecting a network among all the ones with realistic characteristics. Because of what just said, the network is called a random network or a random graph.

To obtain random networks with the desired characteristics, several algorithms are at disposal. In the following of this section, we present the most important ones.

The first and simplest model to generate random networks is based on the seminal works of Gilbert (1959) and Erdös and Rényi (1959). The algorithm allows creating a network with an expected number of links given the number of nodes (i.e., with an expected average degree), and it creates a degree distribution that approximates a Poisson distribution. The algorithm is as follows: create N nodes, then consider any possible link between nodes, and create it according to a unique and constant probability value p.

The expected number of links in the network can be computed by multiplying the p value with the number of possible links, which is equal to $(N(N-1))/2$ in an undirected graph. The resulting network shows two important features. First, if the number of links is large enough but without being too large, the network has a small diameter relative to the size of the network. Second, the average clustering coefficient is low.

The second basic algorithm to consider is the one generating small world networks (Watts and Strogatz 1998). Small world networks are typologies of social networks that can be found in several social contexts and that have, at the same time, a small diameter, large clustering coefficients, and some variance in nodal degrees.

The algorithm starts from the creation of a list of the N nodes in random order. The list actually has no boundaries, meaning that the end connects with the beginning so that each node in the list has the same number of nodes preceding and following it for the purpose of the algorithm. The list, thus, can be conceived as a circle or as a one-dimensional lattice. Then some links are created between each node and its neighbors. In fact, for each node, it can be considered a neighborhood made by four nodes, the two preceding and the two following the node in the list. The resulting average cluster coefficient, equal to 0.5, of this temporary network is high and independent from N. However, the temporary network has a large diameter in the order of $N/4$ and identical degrees in nodes. Thus, the second part of the algorithm removes an existing link between two nodes, and it rewires one of the just disconnected nodes to another one randomly selected among the ones not belonging to the neighborhood. This latter mechanism is repeated several times until the desired value of network diameter is obtained. If the algorithm is repeated too many times, the resulting network can lose clustering.

The third kind of random network can be obtained with an algorithm usually called the "configuration model" (Bender and Canfield 1978). It is useful when a specific degree distribution has to be replicated.

The algorithm works similarly to a random wheel. We create a randomly ordered list of N nodes, and then we assign to each node the value of expected degree that allows replicating the degree distribution of interest. We then create a second list with cumulated degree values (e.g., if the degree of the first node is 10 and the degree of the second one is 7, in the cumulated list, we will have a value equal to 10 associated with the first node and 17 associated with the second one). Using the cumulated list as a list of probabilities of link creation, we generate two random numbers from a uniform distribution between 0 and the maximum value of cumulated degrees (i.e., the sum of all degrees). We use the two random numbers to select two nodes. In fact, we select the node with the smallest cumulated degree larger than the random number. We then connect the two identified nodes with a link and decrease in the first list the number of connections to be created in each of the two just connected nodes, and we recompute the list of cumulated degrees with the updated list of degrees. The procedure is repeated until all the desired links in the lists are created.

The algorithm allows perfectly replicating any distribution of degrees, but it has the drawback of allowing the creation of self-links and of duplicate links between the same nodes. The solution to these problems is erasing self-links and duplicate ones at the end of the algorithm, but with care. In fact, the resulting degree distribution will not significantly change only if N (i.e., the number of nodes) is large enough.

The fourth and last kind of algorithm to consider provides the capability of creating scale-free random networks, which are networks with a power law degree distribution independent from N.

The algorithm uses two components. First, the network grows in size. It starts from a small set of initial evenly connected nodes m_0, and then nodes are added to the network one by one until N is reached. Second, whenever a node is added to the network, m links are added (with $m \leq m_0$) according to preferential attachment. Preferential attachment is a mechanism determining the

Table 7.4   Models for the generation of random networks.

Model	Parameters	Algorithm	Features
Poisson	Network size (N)	1. Create N nodes	Distribution of degrees: Poisson
	Average degree (value p)	2. Create any possible link with probability p	Diameter: small
			Average clustering coefficient: low
Small world	Network size (N)	1. Create a circle network with N nodes and links to neighbors	Diameter: small
	Diameter	2. Rewire existing links with far nodes until desired diameter	Average clustering coefficient: high
Configuration	Network size (N)	1. Create lists of degrees as desired	Distribution of degrees: as desired
	Degree distribution	2. Use list of cumulated degrees as a random wheel	
		3. Create links and update lists	
Scale-free	Network size (N)	1. Create the initial set of nodes m_0	Distribution of degrees: power law
	Initial set of nodes (m_0)	2. Add nodes creating for each $m < m_0$ links with preferential attachment	Diameter: small
	Links to add for each node (m)		

probability to connect the new node with any other already in the network. The probability of preferential attachment to node i is $k_i / \Sigma_j k_j$, with k indicating the number of links incident to a node. In other words, the probability of a node of receiving new links is determined by the number of links already existing on that node, and more connected nodes receive more new connections.

Table 7.4 synthesizes the four models for generating random networks that have been presented, describing the parameters they require and the features of the resulting network.

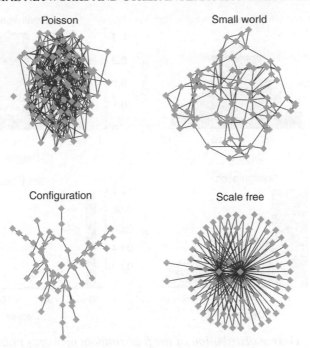

Figure 7.2   Random networks created with the four different algorithms and associated parameters.

Figure 7.2 presents a random network for each one of the algorithms described so far. The network is visualized with the MDS procedure (Corten 2011). In all networks, the number of nodes is 80. In the Poisson random network, $p \approx 0.1$, creating 320 links. In the small world network, the algorithm has rewired 40 links. In the configuration random network, the desired distribution has 40 nodes with a singular degree, 20 nodes with degree equal to 2, 10 nodes with degree 3, and 10 nodes with degree 4. In the scale-free network, m_0 is 4 and m is 3. Figure 7.3 presents the degree distribution associated with each of the four random networks of Figure 7.2.

As final remarks, there are two important points that have to be remembered when using network generation algorithms in ABMs.

First, it is important to remember that they are random networks, meaning random draws among the many possible networks with similar characteristics. It means that random networks have to be treated as other random components of models such as random parameters. Thus, it is often important to use repetitions (i.e., multiple random networks) in order to neutralize the determinism introduced in the simulation by a single random draw.

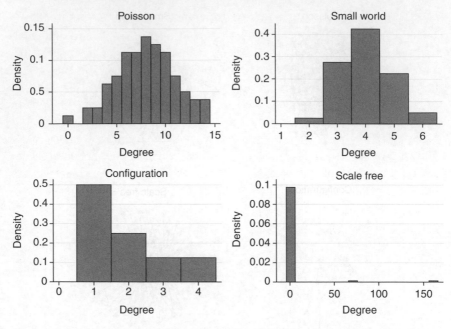

Figure 7.3   Degree distribution of the four random networks presented in Figure 7.2.

Second, it is possible to modify the algorithms introduced earlier in several ways, and there are several other algorithms in the literature (for reference, see Jackson 2008, Newman et al. 2006, Watts 1999), and more will surely be made available in the future. However, the interest of behavioral CSS on network generation models is more on the result than on the process since those models are used to generate realistic interaction structures in ABMs. Thus, more than being interested in the model *per se*, researchers are asked to continuously verify resulting interaction structures. Further, it is important to check the features of the network that is produced even for the purposes of detecting possible coding mistakes and for fine-tuning generation parameters.

7.3   Other kinds of interaction structures

As mentioned in the introduction, social networks are just one of the many possible kinds of interaction structures existing in the real world and that could be modeled in ABM to support the explanation of social phenomena.

The advantage of social networks is that they are an established and self-standing field of research and that consequently there are several analytical tools that support their analysis. When other interaction structures are considered, modelers often recur to *ad hoc* solutions. Among them, it is possible to discern three interaction mechanisms that are particularly diffused and that can be expanded and implemented in several ways. They are interaction through space, with auction mechanisms and with probability scores.

Using space to let social agents interact in ABMs can allow both direct and indirect interaction. Direct interaction happens when, in the model, an agent can travel through space and meet other agents with which to interact. Indirect interaction happens when an agent standing or moving in space modifies the environment. Other agents living in the same space will face an environment that has been modified by that agent. An example of direct interaction in space is Schelling's segregation model (Schelling 1971). Indirect interaction on space can be observed in applications of the Sugarscape model (Epstein and Axtell 1996) and in several ABMs concerning ecology and land use.

In both cases of interaction, space is usually modeled in ABMs as a two-dimensional grid composed by several cells (i.e., a two-dimensional lattice). Interaction can happen between agents that occupy the same cell at the same moment or with spatial neighbors. In developing social ABMs, it is common to refer to two well-known definitions of spatial neighborhoods (Figure 7.4). Von Neumann neighborhood is made by the four nearest cells that are perpendicular to the cell where the agent is. Moore neighborhood is made by all the eight cells surrounding the one where the agent is located. Such kinds of neighborhoods can also be extended if interaction is possible on longer distances.

Most platforms supporting ABMs provide specific programming libraries to manage and visualize space and to allow modeler to easily program movement and interactions on spatial representations. Further, ABM platforms also make available spatial tools based on geographic information systems (GIS) that are particularly useful when the model mimics social dynamics that happen in a specific area for which a GIS map is available (e.g., a case study).

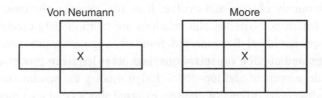

Figure 7.4   Von Neumann and Moore neighborhoods in a two-dimensional grid space.

Auction mechanisms are the second kind of interaction structure frequently found in ABMs. In fact, in particular in economics, auctions are the interaction structure governing several phenomena. They are behind order books in financial markets but also the allocation mechanism of gross markets of several commodities and fresh products. Auctions are either modeled implementing the simple interaction scheme of the game theoretical approach or with more complicated and realistic interaction structures when known (e.g., order books in financial markets).

The third kind of interaction structure to consider is the one based on probability scores. It is used when it is known the profile of partners participating in interactions. Knowing the distribution of the characteristics of the potential partner, a probability score can be computed for each other agent in the model similarly to what is done in statistics with propensity scores. The ABM then will randomly choose interaction partners according to those probabilities.

This method is used in social models relying on survey data that allow modeling both the distribution of characteristics in the population and matching probabilities. Similarly, it is applied in models about the economy and in particular in those focusing on innovation, technological processes, and economic cycles that rely on data about supply chains and production processes such as input–output tables. This method allows direct interaction only.

Finally, as multiple partially overlapping social networks can coexist in the same model (e.g., a network of friends and a network of work colleagues), it is possible to include in ABMs, if needed, any combination of the interaction structures described earlier.

7.4   Critical issues: Time and behavior

The interaction structures described in this chapter have an important limitation. Their definition is static, meaning that it does not evolve over time in the model (though interaction structures can vary over time if the definition is probabilistic as when using probability scores).

In the actual world and because they are social products, interaction structures continuously change and evolve. It is, for instance, the case for social networks, where new friends and relations are continuously created and old acquaintances could get disconnected. Similarly, regulatory processes innovate market rules and auction organizations, and other long-run processes such as economic development and top-down design modify the spatial environment. Besides endogenous drivers of change, external shocks can also modify interaction structures. From that perspective, there can be either long-run processes that codetermine structures' evolution or short-term event that briefly and completely upset interactions. An example of this latter kind of exogenous shock is a disaster that completely impedes interaction along established structures and that could also carry on long-lasting impacts.

Although a static definition is, obviously, not always the one to adopt in behavioral CSS, it still has much value. In fact, several social phenomena that can be studied according to this approach do not require a dynamic interaction structure.

However, whenever a dynamic interaction structure is required, the behavioral CSS approach provides all the needed tools. In fact, when modeling dynamic interaction structures, two elements are critical, time and behavior. First, we need to take care of the different timing of different dynamics, since the evolution of interaction structures goes one at a pace usually longer than the interactions happening along the structure. Secondly, we need to include in our analysis and to model the behavior that changes the interaction structure. This means, for instance, to model the organizational behavior that leads to new market regulations or the individual behavior of preferential attachment that shapes social networks.

Finally, it is also possible to create models specifically aimed at explaining and predicting the dynamics of interaction structures, and such models can be later integrated with others in a cumulative process of ABM development.

The criterion to evaluate if a static interaction structure is sufficient or if a dynamic one is necessary is based on time scales. If the social phenomenon on focus emerges frequently and shortly and its explanatory social mechanisms "lives" in a short time span that allows taking the social structure as given, then the interaction structure can be studied as static. On the contrary, if the phenomenon of interest takes decades to emerge and reproduce, it is undeniable that the evolution of interaction structures should be considered as well.

In summary, dynamic interaction structures contribute to explaining social phenomena similarly to sophisticated agents presented in the preceding chapter. When the evolution of social interactions and of behavior plays an important role in the causality of social phenomena, ABMs should include their dynamics.

8

An example of application

The aim of this chapter is to present an example showing how to apply the methods presented throughout the book. The example shows how to study behavioral data and how to create an agent-based model (ABM).

The ABM is described here along some model results that are aimed at showing some of the potential benefits of the model. The example, in other words, is not aimed at answering a precise research question but at pointing out the steps required to apply the method to an interesting social phenomenon. Many references cited throughout this work and others in the literature (e.g., Boero et al. 2010, Bravo et al. 2012, etc.) are good examples of the results obtainable when using the method to address specific research questions and hypotheses.

The presentation of the model in this chapter is focused on describing its main components, while the technical details of its implementation are provided in the appendix.

Behavioral Computational Social Science, First Edition. Riccardo Boero.
© 2015 John Wiley & Sons, Ltd. Published 2015 by John Wiley & Sons, Ltd.

In this example, we use experimental data as source of information about agents' behavior. In particular, we use an economic experiment that studies a social dilemma. The example, in fact, is about the study of cooperation and free riding among subjects in a game that mimics a standard problem of voluntary public goods provision. It is a tradition of research that has involved many sociologists, psychologists, and economists for decades, and it is still of much interest today (for a literature survey, see Ledyard 1995).

The choice of using an abstract public goods provision framework, as the one implemented in the experiment, is due to the aim of testing the methodology presented and in particular the part concerning the possibility to use experimental data as behavioral ones. Moreover, the model is extended to consider different interaction structures allowing the demonstration of the possibility to investigate behavior and interaction in conjunction and thus social causal mechanisms (Chapter 2).

Because of the application to a specific social dilemma, the first section of this chapter briefly introduces it. The second section then focuses on the original experiment that is used for the analysis of behavioral data.

The third section describes the models of behavioral agents that are included in the ABM and the analyses that have been conducted on experimental data to calibrate and classify behavioral agents. The fourth section presents an example of the use and modeling of more sophisticated agents, and in particular, it presents reinforcement learning as described in Section 6.2.1.

The last section presents some simulation results. Because the model has been developed with didactic intents, its analysis is aimed at surveying advantages given by the approach, and results are organized according to possible research questions that can be addressed.

8.1   The social dilemma

Voluntary public goods provision is a consolidated research theme. It has been formalized and studied using "games": they present a stylized situation that resembles common public goods provision problems like the provision of a television among roommates or of an ambulance in a community.

Games are simpler than real situations; thus, they can be studied with formal tools like the ones provided by game theory to understand the possibility of equilibria, depending on the payoff structure and on subjects' behavior. In fact, a game is composed of some players that have to make some decisions upon their endowment. Depending on the aggregate outcome, they receive a payoff. Exploiting the theoretical approach, scholars try to understand the different kinds of equilibria and how robust such equilibria are depending on the hypotheses on subjects' behavior.

In the case of the standard public goods game, players have an endowment of tokens, and they have to choose the amount to invest in a private good and the amount to invest in a public good (they generally have to consume all their endowment; thus, the choice can be reduced to deciding how much to invest in the public good).

The two goods have a different return rate, higher for the private one, but contributions of all group members (generally the group composition ranges between 4 and 10 people) to the public good are summed, and the revenues of the public good are enjoyed indistinctively by all group members (that is defined by the concept of public good, which is a nonappropriable one).

8.1.1   The theory

Exploiting game theory concepts and its approach characterized by the presence of agents behaving according to the rational choice model, the payoff structure emerging from those return rates creates two important equilibria.

The Nash equilibrium is the one chosen when players behave "fully rational" in the sense of expected utility maximization. It is a robust equilibrium, the most robust because the only one "self-reinforcing" in the sense usually meant in game theory.[1] It is thus the expected result from the point of view of that theory. The theoretical prediction is, in other words, that players do not cooperate at all, they are all free riders, and they invest their entire endowment in the private good.

Another equilibrium is the Pareto one, that is, the one that guarantees the maximization of collective welfare, but it is weak because it is not self-reinforcing. It is represented, in the game, by the case in which all players cooperate at the maximum level possible, that is to say when they invest their endowment completely in the public good.

To better explain why the Nash equilibrium is the one forecasted by game theory, it is worth focusing on its feature of being self-reinforcing: even if a subject knows that the Pareto equilibrium is the best solution possible, he or she will know that each game participant is subject to the temptation of a small deception. Each small deception from the maximum contribution level (that is to say the situation in which even just one token of the total endowment

---

[1] It is a self-reinforcing equilibrium because whatever the choice made by other players (i.e., other group members), the player can improve its monetary payoff by defecting (i.e., contributing all the endowment to the private good).

is invested in the private good) means, *ceteris paribus*, an increase in welfare (i.e., a higher payoff) for the deceptive subject.

Thus, considering that the temptation to deceive is strong and that a fully rational agent should expect that all the other participants are making the same considerations and are subject to the same kind of temptation, the only rational and prudent behavior is to not invest any token in the public good. This choice, in fact, protects against others' possible deceptions. The Nash equilibrium, even if known as a worse solution than the Pareto one, should be the one to be chosen except for the cases in which there is the possibility of "strategic behavior."

In fact, when, for instance, there are repeated interactions in the same group of people for an uncertain number of times, it is possible to exploit behavior as a device to communicate for choosing, as a group, a higher equilibrium[2] (a strategic behavior is thus similar to a signal directed to others). Even if direct communication is not possible and participants do not know each other, subjects can try to build up a reputation to cooperate. In these circumstances, cooperation is perfectly rational even under the strict assumptions of the rational choice model. However, the key condition to make that possible is that subjects do not know, ex ante and during the interaction as well, the number of times they interact.

This last condition influences the possibility of strategic behavior and the expected systemic outcome for the following reason: a subject who knows that this is the last time meeting the other participants knows also that his or her own deception, even if they were cooperating before, means a higher payoff for his or her choice. But if he or she knows that other subjects know the same, it would be necessary to anticipate the choice of deception in the game session before the last. But if others know that as well, it would be necessary to anticipate the free-riding behavior to the session which is twice before the last, and so on. Thus, following this kind of reasoning, which game theorists call "backward induction," a player in a public goods repeated game where the information set contains the duration of the game rationally free rides from the beginning and cannot use strategic behavior.

In summary, game theory, which is the main theoretical reference for the explanation of social dilemmas, predicts complete lack of cooperation unless a long list of conditions is valid (i.e., repeated interaction with the same set of partners and without any possibility to know or estimate the duration of

---

[2] For instance, it is rational to cooperate in iterated prisoner's dilemmas when other people have signaled the exploitation of a tit-for-tat strategy (i.e., a reciprocating behavior—that argument has been developed in Axelrod (1984)). And even if the signal is misunderstood, the decision to cooperate or defect can be used to affect other players' choices in following decisions.

interaction). In general, the theory is entirely based on the assumption that individuals behave like in the rational choice model.

8.1.2 Evidence

In the actual world, there is much evidence of social phenomena similar to the voluntary provision of public goods and where there is much cooperation. It is also very unrealistic to reduce those many phenomena to cases where the conditions discussed earlier are valid. Further, there are much examples of cases where cooperation has worked for long times and then it has stopped. Game theory seems not able to provide much insight on these phenomena. In general, the analytical contribution of game theory is twofold but rather limited. First, it provides a formal setting for the investigation of these social dilemmas. Second, it provides tools to design changes in economic incentives to drive behavior in dilemmas, even if economic incentives are only one piece of the picture. However, it does not provide any analytical contribution to the explanation of the actual phenomenon.

Many economic experiments have been made to test game theoretical hypotheses and results. This vast literature can be seen today as a long-lasting attempt to identify the elements of the theoretical model that require modifications.

The experimental methodology, in particular, allows controlling for nuisance factors and to test the two main components of theoretical models that are game design and assumptions on behavior. All experiments of public goods provision point out, in general and if further very strong economic incentives toward defection are not added, positive levels of cooperation. Experiments, in summary, point out the external validity of game design but the weakness of the rational choice model.

To make a few examples from this vast literature, the experiment we use in the following of this chapter (Andreoni 1995b) shows a contribution to the public good that is, in the first three rounds, greater than 50% and then slightly decreases to the average value of 26.5% of the last round (the 10th one). This experiment has an interaction structure that is a "single shot repeated," which means that each time subjects participate in the game, they know that their group members have been randomly chosen in a draw form a larger set of individuals.

Other similar experiments reported in literature confirm such results, showing that in experiments made by 10 single-shot interactions, subjects start contributing nearly half of their endowment, slightly decreasing it later, until reaching, at the end, a value that ranges between the 15 and 25% (e.g., Andreoni 1988, Isaac and Walker 1988).

8.1.3   Our research agenda

Because experiments confirm positive levels of cooperation in most social dilemmas of voluntary provision of public goods, contemporary research agendas in this field are focused in the search for elements that can improve modeling of behavior. For instance, there is much attention on the role played by social norms and social heuristics (e.g., reciprocation, fairness, etc.) and on the evolution of social interactions (e.g., social networks) where unreliable and defective individuals tend to remain isolated.

From this perspective, it is important to have at disposal models to test hypotheses of causal dynamics and social mechanisms that go beyond the game theoretical ones and that allow flexibility in both behavior and interaction. The model we present here shows how to apply the methodology we have discussed, and it aims, analytically speaking, at providing a formal tool to answer research questions following the contemporary research agendas that have just been mentioned.

8.2   The original experiment

For developing this example, we use behavioral data collected for an experiment presented in Andreoni (1995b) and that has been published on the Internet.

The original experiment was conducted to investigate research hypotheses similar to the ones mentioned earlier. The experimenter in fact used three experimental conditions in the attempt to measure the impact of two possible reasons for the emergence of high cooperation levels. In particular, the experimenter looked for "confusion," which summarizes all the phenomena of misinterpretation of game instructions and payoffs by subjects, and for "kindness," with which are meant altruism, reciprocity, warm glow (Andreoni 1990, 1995a), and so on.

In each one of the three experimental conditions, 40 subjects played the game, making a total of 120 subjects.

The first condition is called "regular," and it is the standard framework for public goods provision experiments. It is a 10 single-shot repeated experiment, in which group composition is randomly changing and the group size is 5. Subjects, at the beginning of each cycle, receive 60 tokens (i.e., the endowment): for each token, they choose to invest in the private good they know they will receive 1 cent of a dollar, and for each invested in the public good, they will receive half a cent.

The return subjects receive from the public good (i.e., 0.5 cents for each invested tokens) is obviously computed over the sum of tokens contributed by all subjects in the same group. In this way, the Nash equilibrium is the one in which all five group members free ride and do not contribute to the public good: each one of them invests his or her entire endowment (60 tokens) to the private good and receives back 60 cents. The Pareto equilibrium, on the contrary, is the one in which each group member invests his or her entire endowment in the public good earning 150 cents (the group contribution to the public good in fact is of $60 \times 5 = 300$ tokens, generating a return of 150 cents per group member).

The "weakness" of the Pareto equilibrium, according to the game theoretical approach to rational choice, is due to the fact that there is a strong incentive toward deception. If a single player chooses to free ride in a cooperative group where all the other members contribute their endowment to the public good, he or she will receive 180 cents (60 cents from his or her investment to the private good and 120 from the others' contribution to the public good), and the other group members will receive only 120 cents.

The regular condition is a sort of control group, and it is the only condition we consider for the rest of our analysis. The original experiment included also a "rank" condition and a mix of the other two called "regrank," but we do not consider them here.

Results of the condition we focus on here observed in the original experiments are synthesized in Table 8.1. They show the average amount of public goods contribution (measured as a percentage over the endowment of 60 tokens), the standard deviation of these values, and the number of free riders (i.e., subjects who do not contribute any token to the public good) in all experimental rounds and reported as percentages of the total (i.e., 40 subjects). It can

Table 8.1   Mean and standard deviation of the percentage of contribution to the public good and number of free riders in the original experiment (as percentage values).

	Experimental rounds									
	1	2	3	4	5	6	7	8	9	10
Mean contribution	0.56	0.60	0.55	0.50	0.48	0.41	0.36	0.35	0.33	0.27
Standard deviation of contribution	0.38	0.36	0.38	0.41	0.38	0.36	0.37	0.37	0.36	0.36
Number of free riders	8	5	7	10	10	12	12	15	14	18

be noted how cooperation decreases over time, without ever reaching zero, and the number of free riders increases over time but stays well below half of the sample.

8.3   Behavioral agents

We develop six different types of behavioral agents based on the different approaches described in Chapter 5.

In particular, we calibrate five decision models, and at last, we use once classification for some behavioral models. Of the five calibrations, four of them are done on a single-level decision function, and the fifth is done with a multilevel decision tree. Of the four single-level decision functions considered for calibration, the first two use a cardinal decision variable, and the other two use different decision variables, and thus, both the data and the model are modified accordingly. Similar modifications are applied for the analysis of multilevel decision trees and of calibrated behavioral models.

8.3.1   Fixed effects model

The first model we consider is extremely simple and puts in relation the amount of tokens to invest in the public good (i.e., the cardinal decision variable, the output of the decision-making process) with results observed in the previous interaction. The model is a population-level one for the estimation of these relationships, but it acknowledges the presence of hetero-geneity in subjects by estimating a "correction" value for each individual. It is, in econometrics terms, a fixed effects regression model.

The model is thus $y_{i,t} = \beta_1 x_{i,t-1} + \beta_2 z_{i,t-1} + f_i + \varepsilon_{i,t}$, where $y_{i,t}$ is the amount of tokens invested in the public good by subject i at time t. $x_{i,t-1}$ is the sum of the amount of tokens invested in the public good by the members of the group the subject i belonged to at time $t-1$ (i.e., in the previous experimental round). $z_{i,t-1}$ is the amount of earning the subject received in the previous round. Finally, f_i is the individual effect and $\varepsilon_{i,t}$ is the error component.

The estimation of this regression model[3] provides results as in Table 8.2. The table reports only a few individual effects for sake of space and a few data about the estimated model on top. Since the model requires observing data in the previous experimental round, the first round of the experiment is not considered, and the number of observations is 360 being 40 subjects interacting 9 rounds.

---

[3] All regressions presented in this work have been estimated using Stata 13 (StataCorp 2013).

Table 8.2   Results of the fixed effects regression model.

Number of observations	360			
R^2	0.818			

Variable	Coefficient	Standard error	t	$P > \lvert t \rvert$
x_{t-1}	0.175	0.026	6.64	0.000
z_{t-1}	−0.175	0.059	−2.97	0.003
f_1	19.845	6.357	3.12	0.002
f_2	48.099	5.765	8.34	0.000
...				

Results show a significant role of the independent variables (i.e., inputs for decision-making) we consider. However, the two variables have the complete opposite impact. In fact, the individual level of contribution to the public good increases if group contribution has increased or if earnings have decreased.

Individual fixed effects are significant too except for 11 subjects (over 40, with $p > 95\%$). In the ABM, we consider all estimated coefficients and also the ones that do not look significant from a statistical viewpoint.

8.3.2   Random coefficients model

Searching for a behavioral model capable to more effectively capture individual heterogeneity and avoiding to recur to individual regressions (in fact, there are in total only 10 observations for each individual), we use a random coefficients regression.

The model is similar to the previous one and equal to $y_{i,t} = a_i + \beta_{1,i} x_{i,t-1} + \beta_{2,i} z_{i,t-1} + \varepsilon_{i,t}$. The independent variables, x_{t-1} and z_{t-1}, are the same as in the previous model. a_i is also similar to f_i since it is a constant value for each individual. Coefficients are, on the contrary, much different and equal to $\beta_{1,i} = \beta_1 + v_i$ with $E(v_i) = 0$.

In summary, the estimation of this model allows to find individual coefficients and constants under some assumptions on their distribution at the population level. Results are presented in Table 8.3, reporting coefficients and constants for only two individuals.

Table 8.3 presents the results for the first and the third subjects because the second subject in the experiment always invested the same amount of tokens in the public good and thus estimation is not possible. Overall, and including that subject, there are six individuals for which estimation is not possible for the same reasons. In the ABM, they are modeled as agents not impacted by

Table 8.3   Results of the random coefficients regression model.

| Number of observations | 360 | | | |
| χ^2 (2 d.f.) | 10.44 | | | |

Variable	Coefficient	Standard error	z	$P > \|z\|$
a_1	6.622	15.048	0.44	0.660
$x_{1,t-1}$	−0.019	0.956	−0.20	0.842
$z_{1,t-1}$	0.206	0.212	0.97	0.332
a_3	23.909	15.649	1.53	0.127
$x_{3,t-1}$	0.213	0.091	2.34	0.019
$z_{3,t-1}$	−0.394	0.178	−2.21	0.027
…				

independent variables (i.e., their coefficients are set to 0) and with a constant value equal to the amount of tokens invested. Further, from a statistical viewpoint, of the 102 coefficients estimated (3 coefficients for 34 subjects), 71 of them are not significant ($p > 95\%$), but they are all considered in the ABM.

8.3.3   First differences model

We take into consideration another way to deal with the complex nexus between regularity and heterogeneity in behavior by referring to a first differences model. From an analytical viewpoint, it could seem to be very similar to the fixed effects one because it assumes a common model of behavior that works in individuals that have a different starting point. In fact, a first differences model describes how the decision variable should be varied over time, not its absolute value.

The model is $y_{i,t} - y_{i,t-1} = a + \beta_1 \left(x_{i,t-1} - x_{i,t-2} \right) + \beta_2 \left(z_{i,t-1} - z_{i,t-2} \right) + \varepsilon_{i,t}$, where y, x, z, and ε mean the same as in previous models. However, the β coefficients are now computed over the difference in independent variables observed in the previous round. Because of this latter issue, we need to exclude the first two rounds of the experiment from the regression dataset (i.e., we need to use the first two rounds to compute the values of independent variables used in the third round).

Results are presented in Table 8.4, and estimations seem statistically significant, but the aggregated model seems performing poorly (see the R^2).

In summary, this behavioral model does not directly decide the amount of investment in the public good but how to modify the preexisting level of this variable. The ABM has thus to implement also the application of the change. Further, this behavioral agent needs differences in independent variables for decision-making, and thus, the ABM computes and updates such values.

Table 8.4   Results of the first differences regression model.

Number of observations	320			
R^2	0.197			

Variable	Coefficient	Standard error	t	$P > |t|$
a	-3.093	0.990	-3.12	0.002
$x_{t-1} - x_{t-2}$	-0.209	0.029	-7.31	0.000
$z_{t-1} - z_{t-2}$	0.468	0.054	8.69	0.000

8.3.4   Ordered probit model with individual dummies

Since investments in the public good are repeated, it is possible to separate decision-making in two distinct phases. In the first, subjects decide if they are satisfied with their previous choices, and if not, they modify their level of investment, either increasing or decreasing it. In the second phase, if they modify the level used in the previous round, they decide the amount of the change.

Focusing on the first phase, we consider that subjects have three possible choices: decreasing the investment, keeping it at the same level, or increasing it. We thus transform the data according to these possibilities, looking for changes in investment levels. The new decision variable takes value 0 if public goods investment decreases, 1 if it remains constant, and 2 otherwise (i.e., if it increases). The new decision variable is an ordinal one. We also apply the same transformation to the two dependent variables, total group contribution to the public good and earnings, that thus take only the values 0, 1, or 2 depending on their dynamics.

We thus study this modified dataset and we consider the possibility to have heterogeneity in subjects. We use an ordered probit model to deal with the nonlinearity of the ordinal dependent variable, and we use individual dummy variables to take into account heterogeneity.

The model is thus $y_{i,t}^* = a_i + \beta_1 x_{i,t-1}^* + \beta_{2,i} z_{i,t-1}^* + \varepsilon_{i,t}$ where the star in independent and dependent variables indicates that they are the new ones transformed as just described. Even in this model, the first two rounds of the experiment cannot be used for the regression since they are required to compute lagged values.

The estimation of the model provides results as in Table 8.5 where most individual dummies are not reported for matter of space. Of the 40 individual dummies, 21 of them are not statistically significant ($p > 95\%$), but they are used in the ABM.

It is important noticing the last two rows of the table, where cut-points are presented. They are values used to transform the model in a nonlinear one.

Table 8.5   Results of the ordered probit regression model.

| Number of observations | 320 | | | |
| χ^2 (41 d.f.) | 54.07 | | | |

Variable	Coefficient	Standard error	z	$P > \lvert z \rvert$
x_{t-1}^*	−0.289	0.088	−3.28	0.001
z_{t-1}^*	0.401	0.086	4.65	0.000
a_1	−1.198	0.609	−1.97	0.049
a_2	−1.060	0.587	−1.81	0.071
...				
First cut-point	−1.585	0.456		
Second cut-point	−0.310	0.453		

In fact, the equation written above is strictly linear, and using just the estimated coefficients allows predicting a cardinal variable, not an ordinal one as intended. Cut-points are thus thresholds and the predicted value of the dependent variable is 0 if the computation of the equation above with estimated coefficients gives a value smaller than or equal to the first cut-point. The dependent variable is 1 if the value is greater that the first cut-point and smaller than or equal to the second cut-point. Otherwise (i.e., if the linear value is greater than the second cut-point), the dependent variable takes the value 2.

Behavioral agents calibrated with this technique include a further and final algorithm in the ABM. In fact, the output provided by the behavioral model is the decision of changing the level of investment in the public good and the eventual direction of the change. In order to transform such decisions to precise values of investment, we use the distribution of changes observed in the experiment for each individual. In other words, we analyze each subject in the experiment, and for each one of them, we compile two lists, one containing the positive changes in the level of contribution to the public good and the other containing the negative changes of the same variable. Then, for sake of simplicity, we assume that the values contained in each one of those lists are normally distributed, and thus, we consider their mean value and their standard deviation. In the ABM, we use the mean and the standard deviation of the distributions to compute the precise values of changes. Thus, for instance, if a subject decides to increase the amount of investment in the public good, we compute the precise value of such a change by means of a random draw from a normal distribution parameterized over the increases of contribution observed in the experiment for that subject.

8.3.5   Multilevel decision trees

The last calibrated behavioral agent we consider is a multilevel model estimated with the technique called symbolic regression or genetic programming (GP; see Section 5.2.2).

By looking for multilevel models of behavior, we are searching for a representation and explanation of behavior in which we impose less constraints than in previous models. In fact, though being a calibration activity, we do not impose on the model any predetermined structure a priori. The calibration procedure tells us how many levels are used by subjects in decision-making. Further, we do not impose a functional representation of behavior, and we include the possibility to have conditional statements (e.g., if a condition is valid, choose something) that are impossible to represent with a single mathematical function.

A further assumption we consider, as in all behavioral agents, is that behavior is stable, meaning that the causal relationships that determine it do not change over the considered interactions.

Before applying the GP heuristic, we need to perform two preliminary activities. First, we need to transform the observed data. Second, we need to define the alphabet that is used in decision trees. The two activities are not independent, and in conjunction, they define the space of possible solutions for the optimization heuristic that is to say the possible shapes that decision trees will have.

Talking about data transformations, we know for sure the kind of choice subjects are facing. Further, we know they all have the same information about the experiment rules, they all have information about the results of their last round, and they are facing the choice about how many tokens to invest in the public good. In other words, we are sure that all subjects have at disposal the same information, and thus, we know that eventual heterogeneity cannot be modeled a priori but can only emerge in the calibration process (as assumed also in previous behavioral agents considering heterogeneity). Moreover, for each player, and for each round, we know his or her available information, that is to say, for instance, that we know that player number 17, when choosing the contribution level in round 6, knows that in round 5 he or she has contributed 10 tokens to the public good, that the total contribution of the group he or she belonged to was of 30 tokens, and that his or her payoff was of 65 cents.

Subjects decide an output informed by some inputs, but they also have memory. They thus can remember the information they received in a few previous interactions, as implicitly assumed also in the first differences and ordered probit models described earlier. With the analysis used here, we define the maximum amount of memory that we can expect in subjects, but it is the calibration technique that finds how much memory has been used.

The transformation of the original data has been applied to both inputs (i.e., the information received on each round and the one kept in memory) and outputs (i.e., the decision taken), and it has been done with the following function:

$$f(x_{i,t}) = \begin{cases} 0 \text{ when } x_{i,t} > x_{i,t-1} \\ 1 \text{ when } x_{i,t} = x_{i,t-1} \\ 2 \text{ when } x_{i,t} < x_{i,t-1} \end{cases}$$

The variables that are transformed with this function are the output, which is the investment in the public good, and the inputs, which are the total contribution to the public good in the group and the received individual earnings. The meaning of the transformation function is immediate. A value, for instance, the one of earnings, at time t is compared with the one at time $t-1$, and if the former is greater, the new data value is 0, if they are equal it is 1, or 2 for the remaining case. Such a transformation works of course because of the "assumption" about the presence of memory in human subjects.

Further, in comparison with the two last behavioral agents presented earlier, we extend memory to the possibility of recalling two past transformed values of each source of information, which equals remembering three precise values for those variables. The first three rounds of the experiment are thus not directly considered for calibration.

As an example, in Table 8.6, inputs and outputs of a player at round 5 are shown. Elements are listed in the first column and their precise value on the second. In the third column, the relationship between the values considered is written, and then in the last column, the transformed value is reported, which is the elaborated information available for making a decision. In the table, pg_t means the player's contribution to the public good at time t, gr_t the group total contribution at time t, and ea_t the monetary payoff obtained at time t.

In order to set up the GP, we also need to define its alphabet that is made by functions and by terminals (or terminal nodes). In other words, thinking about the fact that we are calibrating decision trees, we need to specify what can appear in the nodes of those trees. We use as terminals all the possible values of inputs and outputs, and they thus can take the value 0, 1, or 2. The decision tree can thus refer to a source of information by calling it (in other words, it is possible to have a rule as "increase, decrease or keep constant the public good investment directly following what has recently happened for

Table 8.6   Example of inputs and outputs available at time 5.

	Original	Relation	Transformed
Input			
Public good contribution $t-1$	$pg_4 = 30$	$pg_4 < pg_3$	2
	$pg_3 = 45$		
Public good contribution $t-2$	$pg_3 = 45$	$pg_3 > pg_2$	0
	$pg_2 = 0$		
Group investment $t-1$	$gr_4 = 170$	$gr_4 = gr_3$	1
	$gr_3 = 170$		
Group investment $t-2$	$gr_3 = 170$	$gr_3 > gr_2$	0
	$gr_2 = 150$		
Earnings $t-1$	$ea_4 = 115$	$ea_4 > ea_3$	0
	$ea_3 = 100$		
Earnings $t-2$	$ea_3 = 100$	$ea_3 < ea_2$	2
	$ea_2 = 105$		
Output			
Investment in the public good	$pg_5 = 30$	$pg_5 = pg_4$	1
	$pg_4 = 30$		

the group total investment," that is to say directly referring to the information contained in the input variable called "group total investment").

Functions are Boolean operators (AND, OR, NOT) and conditional operators (IF, THEN, ELSE), so that it could be possible to find rules such as "IF earnings are growing AND group contribution is decreasing, THEN increase contribution, ELSE keep it constant."

Techniques for calibrating decision trees such as GP are characterized by the need to face problems such as identification (i.e., multiple solutions; Rieskamp and Hoffrage 1999) and data overfitting (in particular because of scarcity of data). Further, we are interested in calibrating behavioral agents with a high degree of realism and of cognitive plausibility. Because of these reasons and to deal with these requirements, we elaborate the fitness function used in the GP.

In fact, the simplest possible version of the fitness function would be the one just considering the representation of experimental data so that possible solutions are evaluated on their capability to replicate what observed in the lab. The GP would simply minimize the error in predicting experimental data. However, we add to this fitness function a small award to shorter solutions. In other words, the fitness function still minimizes the error in replicating experimental data, but if more than one solution provides the same degree of replication accuracy, the one that has the simpler tree is preferred. The simplicity of a decision tree is computed as its number of nodes.

The GP has been executed choosing several more parameters, listed in Table 8.7. The parameters have been chosen with an iterative process of trial and in order to allow a single run of the inference process being effectively executed individually for each subject. Using 50 generations of 500 candidate solutions, the GP generates a possible solution to the problem represented by the best performing decision rule present at the end of the 50th generation.

To have more confidence about the outcomes, for each subject, 20 calibration runs have been made, each with a different random seed. In fact, because GP follows a learning path that is dependent on the starting point and such a point is randomly chosen at the moment of setup, the use of runs with different random seeds helps to increase the researcher's confidence about having found the "best" decision rule.

Table 8.7   Main GP parameters and values.

Parameter	Value
Population size	500
Maximum number of generations to run	51
Probability of crossover	90%
Probability of reproduction	10%
Probability of choosing internal points for crossover	90%
Maximum size of trees created	No
Maximum size for trees created in the initial population	6
Probability of mutation	0.0%
Probability of permutation	0.0%
Frequency of editing	0.0%
Probability of encapsulation	0.0%
Condition for decimation	No
Decimation target percentage	0.0%
Generative method for initial random population	Ramped half-and-half
Basic selection method	Fitness proportionate
Spousal selection method	Fitness proportionate
Fitness type	Adjusted fitness

Executed runs have shown some dissimilarity. In some cases, the execution of 20 induction processes with different starting random seeds has been superfluous, because they all evolve the same solution. In other cases, it has demonstrated the need for such a procedure, because in some runs, a clearly better (in the sense that it has a higher value of fitness) solution has been found than in other runs. In some cases, different solutions with the same value of fitness have been found: they generally have a similar structure, and they mostly differ because of the source of information considered to lead to the choice. In such cases, there are not criteria of plausibility to guide the selection of the decision rule (Boero 2007), and for the ABM, one of them has been randomly chosen among the possible ones.

The result of the inference process has been the obtainment of 40 rules of behavior, one for each subject. Of the 40 rules obtained, 34 of them perfectly replicate all observed decisions (i.e., seven decisions because the first three experimental rounds cannot be used). The other six rules fail to replicate one decision and effectively represent six decisions.

Resulting decision rules can be depicted as decision trees, as in Figure 8.1 that is the one followed by player number 40. The same decision rule can be also showed as in the "pseudocode" that follows:

IF (GROUP CONTRIBUTION (T – 1) DECREASED)

THEN FOLLOW EARNING (T – 1)

ELSE DECREASE CONTRIBUTION

In the ABM, they are simply transformed in the programming language used by the ABM platform adopted (see the Appendix). The decision variable is then transformed in precise values of investment in the public good by

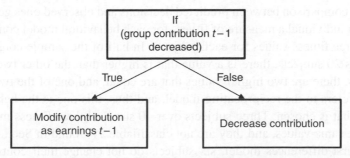

Figure 8.1   Behavioral rule of player 40 represented as a decision tree.

randomly drawing the change value from a normal distribution using the mean and standard deviation of contribution changes observed in the experiment for each subject (as in the ordered probit model).

8.3.6   Classified heuristics

There is a vast literature in the field of behavioral economics and evolutionary economics that has studied social dilemmas similar to the one considered here and that has pointed out the importance of the role played by social heuristics such as reciprocation or conditional cooperation. Such social heuristics are motivated by social norms of fairness but also by evolutionary drivers that award the establishment of cooperative dynamics.

We thus consider the possibility to have a reciprocating behavior that cooperates conditionally on the presence of personal earnings. The first behavioral model we consider is thus a behavior that varies the level of contribution according to the changes observed in the level of earnings. In other words and as an example, if a subject observes his or her personal earnings decrease, he or she will decrease his or her contribution in the following interaction.

Another possible behavioral model we consider is free riding. Subjects behaving like that tend to decrease their contribution level whatever information they receive.

The third model of behavior we consider is the altruistic players or unconditional cooperators. They tend to increase their level of contribution whatever information they have at disposal (and up to the maximum level possible).

We classify experimental subjects on these three models of behavior as described in Section 5.3. In particular, for each individual, we compare the decisions predicted by the three models with the ones observed in the experiment. For classification purposes, we obviously use a modified dataset that considers changes in contribution levels and earnings such as the ones used in the previous two behavioral agents.

The comparison between predicted decisions and observed ones generates for each individual a measure of fitness for each behavioral model considered (thus three fitness values for each subject). In half of the sample considered, which is 20 subjects, there is a value that is higher than the other two. In 11 subjects, there are two higher values that are equal, and one of the two equal values refers to the reciprocating model, and thus, this one is the behavioral model that is chosen. Three subjects over 40 show all three fitness measures with the same values, and they are not classifiable. As found in Section 8.3.3 in the first differences model, six subjects do not change their contribution level during the experiment.

In summary, the classification leads to the modeling of 3 random subjects (i.e., the ones unclassifiable because with equal fitness values for each heuristics), 6 subjects with constant levels of contribution, 4 unconditional cooperators, 21 conditional cooperators, and 12 selfish players (i.e., free riders).

As in previous models of agents, in the ABM, decisions to vary the level of contribution are transformed in precise values of contribution using distributions of changes of that variables observed in the experiment.

8.4   Learning agents

As final model of behavior in agents and as example of a more sophisticated one, we implement an algorithm of reinforcement learning. In particular, we use the first reinforcement learning algorithm presented in Section 6.2.1.

The possible "strategies" explored by agents are the three models of behavior considered in classified heuristics, which are unconditional cooperation, conditional cooperation, and free riding. Agents behave randomly during the first three simulation steps in order to explore the space of solutions. They keep a separate record of the earnings received whenever they make a decision according to one of their strategies (i.e., they record earnings cumulated because of different behavior), and they use those aggregated values as probabilities to define future behavior.

Similarly to what implemented with several behavioral agents presented earlier, in the ABM, decisions are transformed in precise values of contribution according to the distribution of changes in contribution level observed in the experiment for each subject.

8.5   Interaction structures

We consider four kinds of interaction structures in the ABM. Three of them are modeled similarly to the interactions used in games of voluntary provision of public goods, and they consider fully connected groups. The last interaction structure considers a social network structure such as those discussed in Chapter 7.

The first interaction structure is the one observed in the original experiment. In fact, in the experiment, subjects were randomly assigned to groups on each round, and group composition was recorded. We implement in the ABM the same group assignment, and because in the ABM we can extend interaction beyond the 10 rounds used in the experiment, we use the last group composition observed as the one used in simulation steps following the 10th one.

The second interaction structure is similar. Random groups are randomly assigned at the beginning of the simulation, and the assignment is stable over the simulation. The third interaction structure is the same, but groups are randomly defined on each simulation step, and thus, their composition changes continuously.

These two latter interaction structures are governed also by a further parameter. The group size, in fact, can take the value used in the experiment (i.e., 5) or 2, 4, 10, and 20. The possible group sizes have been chosen to allow the computation of separate groups and the use of all agents (thus, all agents participate to one and only one group).

While the first three interaction structures create separate subnetwork of fully connected and equal groups, the last interaction structure we consider uses a small world network connecting all agents.

The algorithm is like the one described in Section 7.2 (Watts and Strogatz 1998). It starts by creating for each agent a neighborhood of four other agents, and then it rewires 30 links over longer distances. The small world network is randomly defined at the setup of the simulation and then kept constant.

The last issue to discuss about interaction structures is simulation duration. Researchers can choose any discrete value they like, extending the duration of interaction beyond what observed in the original experiment.

8.6   Results: Answers to a few research questions

As said in the introduction of this chapter, we develop and use this ABM with a didactic purpose and without any intent of fully explaining a specific social phenomenon. However, we show in the following some applications of the model to a few research questions selected among the many that the model allows studying. The aim is to provide the reader with the intuition of the several potential uses and benefits of the method.

In particular, we start by replicating the original experiment in order to compare the outcomes generated by adopting different behavioral models. Once the replication of the experiment is verified and the ABM is temporarily validated, we investigate further brief hypotheses concerning both prediction and causality.

Because we intend to compare the performance of the different models of agents' behavior presented earlier, we use simulations where all behavioral agents act like in the experiment during the first three simulation steps. Then, at the fourth step, they start behaving according to their model of decision-making. This is needed because some behavioral agents need to cumulate information before being capable to use their decision-making model. The more sophisticated model of decision-making that is based on reinforcement learning uses random choices in the first three simulation steps.

The use of models in the following text is based on the selection of different parameters. The configurations of parameters that are explored are provided in the electronic file of the ABM distributed on the webpages supporting this book.

Because almost all ABM instances use the generation of random numbers to deal with probabilistic components (e.g., random interaction structures, random endowments, etc.), we present the following results as average ones observed over the population of agents and over 10 repetitions of the simulation, each one run with a different seed for random number generation. However, for a more scientifically robust analysis, the number of repetitions should be increased and be dependent on the amount of probabilistic components in models. We consider here a simpler analysis due to our didactic purposes.

8.6.1   Are all models of agents capable of replicating the experiment?

The first question we can address is whether all the models of behavior in agent we have developed perform "enough well" to replicate, in simulations, what has been observed in the experiment. The behavioral analyses we conducted earlier tell us that the models seem good if taken at the individual level only (but we do not know about reinforcement learning). We now want to check the performance if we include interactions too.

To investigate this question, we set up the ABM as follows. First, obviously, we run several simulations varying the behavior used by all agents without "mixing" different kinds of agents in the same simulation.

Second, we consider in all parameters the values used in the experiment, and thus, for instance, the returns paid by investment in private and public goods are the same as in the experiment (0.01 and 0.005, respectively).

Third, we consider the size of groups that was used in the experiment (i.e., five members), and we assign agents to groups exactly as in the experiment.

Comparing results for simulations lasting as the experiment (i.e., 10 simulation steps) provides results as in Figures 8.2 and 8.3.

Figure 8.2 in particular presents the average level of investment in the public good (or contribution) computed as the mean in the population and over 10 simulations run each with a different random seed.

Results observed in the original experiment are reported in the top right corner, and they show a decrease in contribution that does not converge to

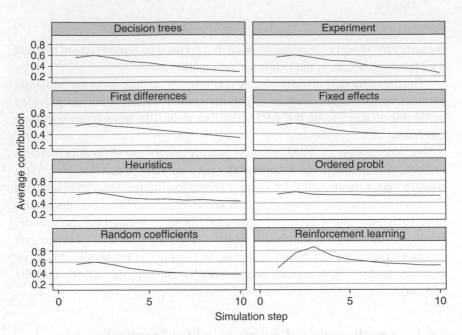

Figure 8.2   Average contribution to the public good in the original experiment and in simulations with different types of agents.

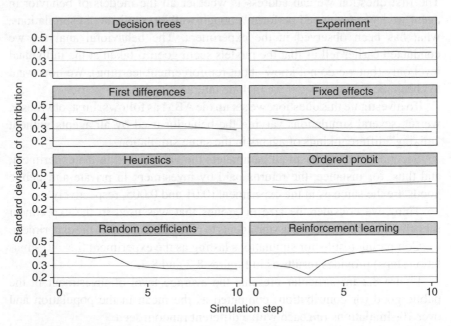

Figure 8.3   Standard deviation of contributions to the public good in the original experiment and in simulations with different types of agents.

complete free riding, similarly as found in several other replications of this game in experimental settings.

All models of agents seem to replicate such stylized fact with more or less precision except for agents behaving with the ordered probit model. Even agents with reinforcement learning replicate the dynamics after the first three steps that are randomly chosen.

Similarly, in Figure 8.3, we study another stylized fact observed in the experiment that is the standard deviation of the levels of contribution. In other words, we also look, partially, at the distribution of contribution values.

In the experiment, the standard deviation remained almost constant. In simulations, results can be different depending on the agents that have been used.

For instance, agents with reinforcement learning and, for a smaller degree, agents using classified heuristics tend to increase differences in contribution levels, probably because the reduction to three behavioral patterns makes behavior converge toward extreme values of contribution. These dynamics are obviously not noticeable if looking at average values only. Agents based on ordered probit calibration and modeling show similar poor results probably because of the same reasons.

Simple calibrated models of agents such as fixed effects, random coefficients, and first differences show the opposite dynamics though decreases in standard deviation are smaller. These models, though partially taking into consideration heterogeneity in subjects, tend to converge toward similar levels of contribution.

Heterogeneous decision trees calibrated with GP show constant levels of standard deviation.

In summary, except probably for the ordered probit agents, all models of behavior seem to replicate well enough what observed in the experiment. However, we cannot conclude that they are validated or effective for all possible uses of the ABM. They are, inevitably, approximation of actual behavior, and thus, they have to be validated in conjunction with the rest of the ABM depending on the use that is done of the model. The mere replication of the experiment is however a good starting point for further uses and analysis of the ABM.

8.6.2   Was the experiment influenced by chance?

We already know that results observed in the original experiment have been replicated several times in similar experiments reported in the literature. We could wonder, however, whether there is a formal way to investigate the external validity of experimental results.

In particular, we can focus on one specific aspect of the experiment that could have influenced its dynamics. In fact, during the experiment, subjects were assigned to groups randomly, but obviously, such random assignment has been observed only once (Andreoni and Croson 2008). It could have been the case, for instance, that in a few rounds of the experiment, all subjects less prone to cooperation were grouped with reciprocating subjects, determining the decreasing levels of contribution. In other words, group assignment, which has been observed only once and which cannot be neutralized in the experiment without recurring to further replications in the lab, could have had consequences on results.

We investigate simulations with parameters as the ones discussed for the replication of experimental results, but we modify the interaction structure so that groups are still fully connected and with five members, but with different randomly chosen compositions of subjects that change every simulation step.

We present results in Figures 8.4 and 8.5 along the ones observed in the experiment (top right corner of figures).

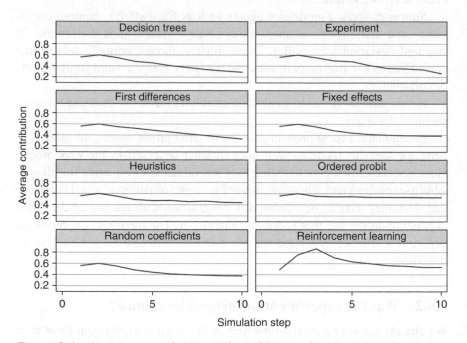

Figure 8.4   Average contribution to the public good in the original experiment and in simulations with different types of agents and random group assignment.

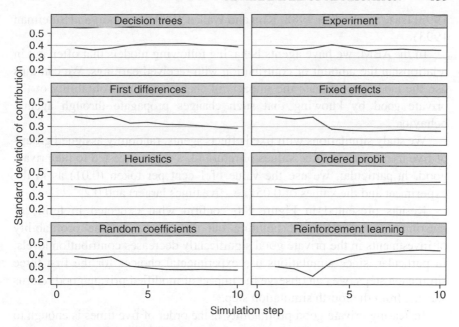

Figure 8.5   Standard deviation of contribution to the public good in the original experiment and in simulations with different types of agents and random group assignment.

Simulations confirm what already observed in the literature in experimental economics. In fact, both average contribution levels (Figure 8.4) and standard deviation of contribution levels (Figure 8.5) neither differ much from experimental ones nor with the results obtained with the original group composition (Figures 8.2 and 8.3).

8.6.3   Do economic incentives work?

Another kind of questions regards the possibility to estimate the effects of changes in parameters. For instance, we can verify whether making defections more tempting modifies contribution levels and quantify the impact.

We know that in social dilemmas, economic incentives define the problem and the payoffs subjects receive. We also know, from the view-point of reasoning, that an increase in the return rate paid to investments in the private good should decrease the amount of contribution and cooperation because it makes defection more profitable. We also have at disposal experiments showing similar patterns (e.g., Brown-Kruse and Hummels

1993, Isaac and Walker 1988, Kim and Walker 1984, Rapoport and Suleiman 1993).

In the ABM, we have agents behaving following models that often put in relationship the amount of contribution with received earnings. We can thus use the model to estimate the impacts of changes in the profitability of the private good by knowing that such changes propagate through agents' behavior.

We study simulations with fixed effects agents, randomly assigned groups of five members and three values of return for a token invested to the private good. In particular, we use the value of 1 cent per token (0.01) as in the experiment and the values of 0.05 (i.e., five times larger) and 0.2.

Results presented in Figure 8.6 confirm what expected in terms of reasoning and plausibility, pointing out that increasing the profitability of investments in the private good significantly decreases contribution levels. In particular, since simulations use experimental choices for the first three simulation steps, we can observe the impact of modified private good returns starting from the fourth simulation step.

Increasing private good profitability in the order of five times is enough to obtain in a few steps the disappearance of cooperation. Larger increases in that

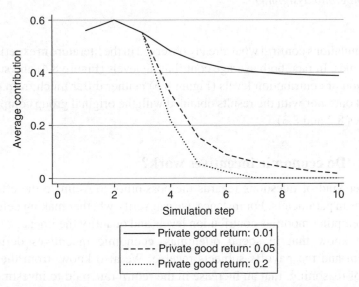

Figure 8.6   Average contribution to the public good in simulations with fixed effects behavioral agents, random group assignment, and different levels of return on investments in the private good.

parameter (and thus larger economic incentives to defect) make that process quicker. With the ABM, we can investigate also other values of parameters, and we can quantify impacts accordingly.

8.6.4 Why does increasing group size generate more cooperation?

Referring again to the literature in experimental economics (e.g., Bagnoli and McKee 1991, Chamberlin 1974, 33 and Walker 1988, Isaac et al. 1994, Marwell and Ames 1979), we know that it has been observed in experiments that increasing the size of groups most of the times implies obtaining higher levels of cooperation.

We can use the ABM this time to explain that phenomenon. In particular, we replicate it with a set of simulations using agents behaving with the model calibrated on random coefficients and all values of parameters as in the experiment except for group size. In fact, we use randomly drawn groups changing every simulation step, but we compare three sizes of groups, which are 2, 5, and 10.

Results presented in Figure 8.7 show that the relationship between group size and levels of cooperation is replicated by the model. By inspecting the

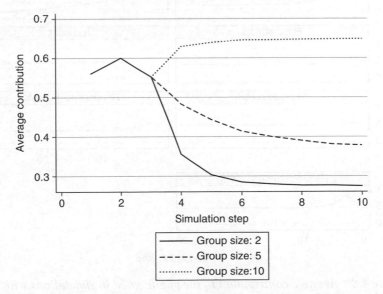

Figure 8.7   Average contribution to the public good in simulations with random coefficients behavioral agents, random group assignment, and different group sizes.

model, we also find evidence of the trivial reason why such a relationship exists. In fact, we can inspect average earnings (not reported here), and we can also observe that for the same levels of investment in the public good, the mere fact of having larger groups means larger earnings because of the pooling of more resources in the public good.

8.6.5   What happens with longer interaction?

We could also wonder whether the ABM is capable to replicate another fact observed in the experimental literature on public goods provision that is the fact that in experiments with longer duration, contribution levels continue to decrease asymptotically to values that, however, are still positive.

In order to verify this feature, we run simulation setup as in Section 8.6.2, with all kinds of agents and randomly assigned groups of five members that change every simulation step.

Results presented in Figure 8.8 confirm the goodness of many of our agents' models even for considering longer interactions. The only two models

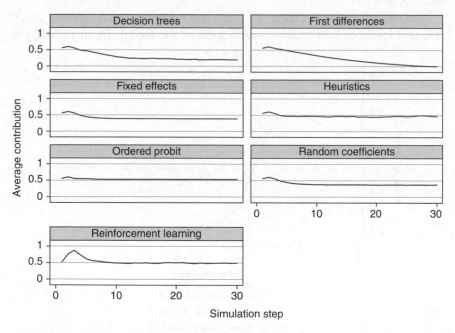

Figure 8.8   Average contribution to the public good in simulations with different types of agents, random group assignment, and 30 periods of interaction.

that present some criticalities are the one calibrated on ordered probit and the one on first differences. The ordered probit model failed already in the replication of the experiment with a shorter interaction.

The first differences model is calibrated to replicate changes (decreases) in contribution levels as observed in the experiment. If the interaction is extended, it continues to apply the same changes (decreases), and thus, it converges toward zero cooperation, which is an outcome very rare in real and experimental settings.

It is also interesting to note that the sophisticated model of reinforcement learning, which is neither calibrated nor classified, generated average cooperation levels about 50%, similarly to the classified heuristics.

8.6.6   Does a realistic social network promote cooperation?

As a final example of ABM application, we can start addressing a broad research question that starts from acknowledging the fact that several real social networks have features similar to small world networks.

The research question is thus whether small world networks promote more cooperation than the fully dense but isolated groups usually modeled in experiments of voluntary public goods provision.

In order to answer this question, we compare simulations where the parameters are generally equal to the experiment but where interaction structure is modeled either as a small world network or as random groups of five fully connected members. Both social interaction structures are randomly drawn at the beginning of the simulation and then kept constant over simulation time. The comparison is particularly meaningful since in fully connected groups of five members, each agent has a degree equal to 4 and to the average degree present in the small world network.

We compare simulations based on behavioral agents calibrated with fixed effects and agents with reinforcement learning.

As Figure 8.9 points out, our superficial analysis suggests that the realistic interaction structure, *ceteris paribus*, contributes to generate slightly higher cooperation levels. With reinforcement learning, the positive contribution is larger, and it emerges over time. In fixed effects agents, the contribution is small but constant.

In summary, the mere fact of having uneven and realistic relationships between agents seems to support cooperation, and the ABM can be used to understand and test the reasons for that and the conditions that make this result possible.

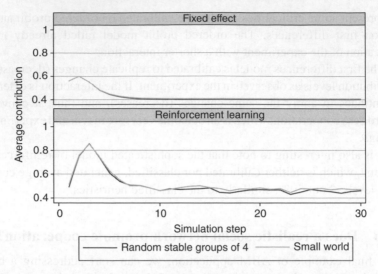

Figure 8.9   Average contribution to the public good in simulations with fixed effects behavioral agents and reinforcement learning agents, over 30 periods of interaction; comparison of two stable random interaction structures, fully connected groups of five agents and small world networks.

8.7   Conclusions

The applications of the model presented earlier outline the analytical power of ABMs built following the behavioral CSS approach. Their presentation and use have been limited and didactic, but they still point out how many research directions can be pursued with such a powerful formal tool at disposal. They are a starting point and a promising one.

The research questions that have been sketched require further analyses that are not developed here. Further experiments, deeper analyses of model results, modifications of the ABM, and extended surveys of the relevant literature could positively contribute in transforming the intuitions we have presented in sound and interesting explanations of important social phenomena with relevant consequences for social welfare.

For instance and referring to the last use of the ABM presented earlier, it could be possible to extend the model and the analysis in order to understand which social network features have impact on the level of cooperation and how those features are in relationship with the decision-making of agents (Boero 2007).

Similarly, it is possible to introduce in the model the possibility to observe the evolution and shaping of the structure of interaction, for instance, allowing agents to enter groups with higher degrees of cooperation or to establish relationships with particularly cooperative partners. Such modifications would probably support emerging evidence about the important role that partner selection plays in establishing cooperation (e.g., Bravo et al. 2012, Chiang 2010, Corten 2014, Grimm and Mengel 2009).

Remaining focused on that model application, another possibility could be to parameterize the model to a specific and actual case of the considered social dilemma. The resulting model could easily be used to understand specific dynamics observed in the real world and also to answer important questions for the design of public policies (Boero 2011).

In using formal models such as ABMs, the fact of referring to information about behavior and interaction provides researchers with the possibility to continuously validate models. The presence of limits to validity, as early noted in philosophy of science, is not a problem but an opportunity because it actually allows improving and developing scientific knowledge.

As discussed throughout this work, validation strictly depends on the use of the model, and thus, it is impossible to find the perfect and universal specification of model components, but instead, it is important to focus on the robustness of them. Robustness is a more adequate word because it contextualizes validity to model use.

To make a few examples, obstacles to robustness can be due to "weak" analyses and starting data, as it would be if we use the behavioral data and analyses presented earlier for building an ABM of transactions in financial markets. The resulting model would probably be not meaningful from a scientific perspective because the specification of model components is not enough robust and versatile to effectively model causal dynamics in such a different context.

Similarly, there can be limits to robustness intrinsic to the model, as in possible further uses of the model for studying the impacts of group size on cooperation. We always have to take into consideration the limits imposed by all model components at once. For instance, we can expect to observe "false" results if we increase group size to the population one since at that point (and probably also when getting close to that point), the interaction structure is transformed from one that randomly changes on each interaction step to one that is randomly designed but stable over time.

For all these problems, it is up to the reader deciding whether the error is in the model or in its use. For the purposes of scientific investigations of social phenomena, the discovery of limits and weaknesses is just an important source of information that supports new discoveries and advancements.

In conclusion, the approach we have presented allows searching for causality of social phenomena from the perspective of social mechanisms. Modeling, in fact, requires approximation and reduction of complexity but not in the field of causality. When social systems are studied, researchers cannot reduce behavior and social interactions because they are the fabric of society.

Appendix

Technical guide to the example model

The example model presented in Chapter 8 has been developed in NetLogo (Wilensky 1999). This appendix presents the details of the technical implementation of the model, and it is intended to support readers in learning how to develop their own models with that software. Nevertheless, even readers who prefer other ABM platforms can find the description of the model implementation useful as an example of some of the algorithms presented throughout the book. This presentation neither covers basics of programming nor basics of NetLogo, for which we suggest to reference the many tutorials and examples distributed along the platform.

The presentation of the implementation is focused on the model only. Simulation results presented in Chapter 8 have been collected by means of a further tool provided with NetLogo that is called BehaviorSpace. The tool allows exploring the parameter space and saving results. Exploration of the parameter space is facilitated by the possibility to run parallel simulations.

Behavioral Computational Social Science, First Edition. Riccardo Boero.
© 2015 John Wiley & Sons, Ltd. Published 2015 by John Wiley & Sons, Ltd.

The name of the tool, however, means that it allows exploring the behavior of the model, and thus, it has no relationships with the investigation of agents' behavior.

ABMs in NetLogo are made by two main components, the interface and the model code. The interface is the graphical user interface (GUI) that allows researchers running the simulation and accessing preliminary results while the simulation is running. The code is the part of the ABM that contains the model and thus agents, their behavior, and their interactions. The first section of this appendix shortly introduces the interface of the example. The second section describes the code.

The electronic file of this model can be downloaded from the webpages that support the book.

A.1 The interface

The interface of a model in NetLogo can contain elements belonging to three main categories. There are in fact elements that allow controlling the simulation, elements that allow defining input parameters, and elements that show features of the simulation when running.

The interface of the example model is presented in Figure A.1, and it contains four elements to control the simulation (they are at the top left corner of the figure). They are two buttons that, respectively, allow the setup and

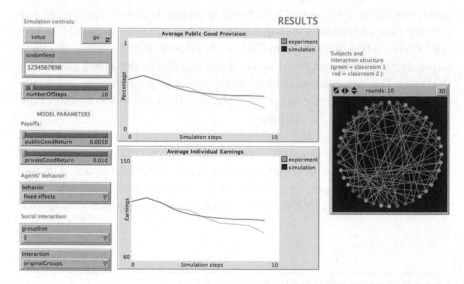

Figure A.1   The interface of the simulation.

execution of the model. Below, there are two inputs that allow setting the seed used by the random numbers generator and the number of simulation steps after which the simulation will stop.

Below simulation controls, there are model parameters. Two of them concern the structure of the social dilemma. In fact, they allow modifying the returns paid to the two different goods (i.e., the public good and the private one) agents invest on. Below, there is an element allowing the selection of agents' behavior from a list made by the possible ones described in Chapter 8. Further below, at the bottom left corner of Figure A.1, there are two controls of the interaction structure. The first one defines the size of groups sharing public goods, and the second one defines the shape of the interaction network. The parameter for the group size is relevant only when the interaction network is based on randomly defined groups, which is the case when members are randomly assigned to groups and groups are fully connected. The group size is ignored when groups defined in the experiment are used because the size was set at that time and when the interaction network is modeled as a small world since in that case group size is heterogeneous and defined by the network.

In the interface, the possibility of selecting options from lists is defined as in Figure A.2 that shows the configuration of the element to select the interaction network. The definition regards the name of the model variable (i.e., in Figure A.2, the "global variable") and the choices that are made available.

Getting back to Figure A.1, the rest of the elements of the interface are used to show model features during the simulation. They are, in other words, plots of some of the results. In particular, there are three elements, two line graphs and a network graph. Starting from the latter, this is a component used in NetLogo to plot interaction space and that shows all agents and their connections. Agents are represented as circles, and their color (green or red) represents whether they were in the first or in the second classroom used in the original experiment.

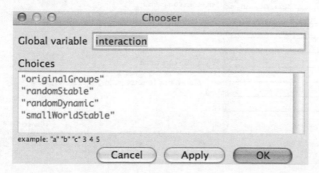

Figure A.2   Definition of input variables in the interface.

The two line graphs present average contribution levels and average earnings in the simulation. For purposes of comparison, they also present the same variables during the original experiment. The definition of graphs is shown in Figure A.3. The model developer has to define titles for the graph and for axes, to set up axes, and finally to define what has to be plotted. For this latter part of graph definition, special objects called "pens" are used. Modifying pen attributes, it is possible to customize the appearance of graphs. The most important attributes to define are the color of the pen and the update command that is the variable to plot. In the case of Figure A.3, the pen refers to the average contribution level to the public good observed in the original experiment, and the update command verifies if the simulation steps are still comparable to the experiment (which lasted 10 rounds), and then it computes the mean value of contribution in turtles (i.e., agents).

In summary, the interface allows the user running the simulation, modifying model parameters, and taking a look at some results. Considering that a scientific use of an ABM requires much exploration of the parameter space and much statistical analysis of results, the interface could seem useless as it does not allow any of those activities. However, that is not entirely true. It is

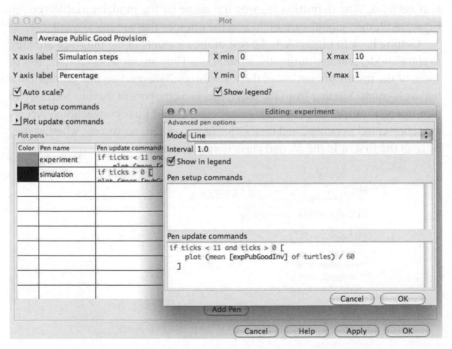

Figure A.3   Definition of graph plots in the interface.

in fact true that interfaces in ABM rarely allow the systematic inquiry of parameters and results, but on the contrary, they are useful when developing and implementing ABMs since they allow researchers immediately controlling simulation results, making easier the identification of design and coding mistakes. Interfaces, further, are also much useful when the model is aimed at supporting the decision-making of stakeholders since they allow such users experimenting with the model easily.

A.2   The code

The "code" tab in the NetLogo platform allows coding the model. The window appears as in Figure A.4. Its main part is reserved for writing the code with useful aids such as automatic indentation and coloring of reserved words. On the top, there is also available the command to find words in the code and a selection button that allows jumping to specific sections of the code (i.e., procedures; see the following).

The code of a NetLogo ABM is made by two main components. They are declarations of variables and procedures modifying variables. The part reserved to variable declarations precedes procedures in the code.

NetLogo follows an object-oriented approach through the use of procedures. In fact, procedures are called by users activating buttons in the interface

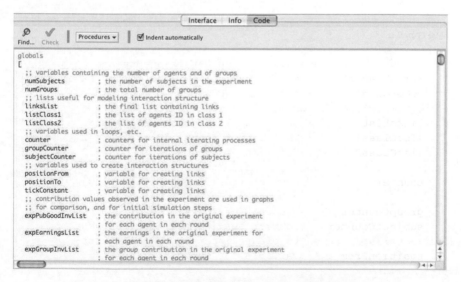

Figure A.4   The window for editing the simulation code.

and by agents and other objects during the simulation. A procedure is thus equivalent to what is usually called a method in object-oriented programming languages.

A.2.1   Variable declaration

At the beginning of the ABM code, there is the part that specifies the variables that are going to be used in the model. If variables have already been referenced in the interface as input parameters, they should not be declared in the code.

Variables can be associated with the model in general ("globals"), with agents ("turtles"), and with links in social networks or with any other object introduced by the modeler. New objects should be defined in this part of the code as well, similarly to variables. The fact that variables are associated with different objects (i.e., the model, agents, links) confirms the object-oriented approach.

The global variables used in the ABM are reported in the following, and they allow managing model parameters (e.g., number of subjects, number of groups, etc.) and procedures that happen at the model level. As examples of variables for managing procedures at the model level, consider the ones related to the definition of the interaction structure (e.g., groups).

Other global variables are used to store individual parameters that are then assigned to agents when created (e.g., lists containing parameters calibrated at the population and individual level in different behavioral models).

```
globals
[
  ;; variables containing the number of agents and of groups
  numSubjects      ; the number of subjects in the experiment
  numGroups        ; the total number of groups
  ;; lists useful for modeling interaction structure
  linksList        ; the final list containing links
  listClass1       ; the list of agents ID in class 1
  listClass2       ; the list of agents ID in class 2
  ;; variables used in loops, etc.
  counter          ; counters for internal iterating
                     processes
  groupCounter     ; counter for iterations of groups
  subjectCounter   ; counter for iterations of subjects
  ;; variables used to create interaction structures
  positionFrom     ; variable for creating links
```

```
  positionTo                 ; variable for creating links
  tickConstant               ; variable for creating links
  ;; contribution values observed in the experiment are used
     in graphs
  ;; for comparison, and for initial simulation steps
  expPubGoodInvList          ; the contribution in the original
                               experiment
                             ; for each agent in each round
  expEarningsList            ; the earnings in the original
                               experiment for
                             ; each agent in each round
  expGroupInvList            ; the group contribution in the
                               original experiment
                             ; for each agent in each round
  ;; parameters values for calibrated models
  fixedEffectsConsList       ; the individual constant
                               coefficient estimated
                             ; with the fixed effects linear
                               regression
  rndCoeffGroupList          ; the individual group total
                               contribution (lag1)
                             ; coefficient estimated with the
                               random coefficients
                             ; linear regression
  rndCoeffEarningsList       ; the individual earnings (lag 1)
                               coefficient
                             ; estimated with the random
                               coefficients linear
                             ; regression
  rndCoeffConsList           ; the individual constant
                               coefficient estimated with
                             ; the random coefficients linear
                               regression
  oprConsList                ; the individual constant
                               coefficient estimated with
                             ; the ordered probit regression
  ;; values of the distribution of changes in
     contribution
  deltaMeanList              ; the list of average values of
                               subjects deltas
  deltaStddvList             ; the list of standard deviation
                               values of subjects
                             ; deltas
]
```

The links used in this model for the interaction structure are extremely simple, undirected, and without weight. Thus, there is no need to add specific variables except for one variable that is used in the procedure for creating small-world networks. Such a variable contains the information on whether the link is one that has been rewired (i.e., a long-distance link) or not (i.e., a link between neighbors).

```
links-own
[
  rewired              ; stores if the link has been rewired
                       (useful for small
                       ; world network creation)
]
```

In NetLogo, agents are called turtles due to its derivation from and adoption of the Logo language. Agents in this ABM need many variables almost all related to agents' behavior. In fact, besides the first variable that is an ID used to associate virtual agents to the individuals in the experiment (for instance, for correctly associating calibrated parameters), all the other variables that are listed in the succeeding are needed by the different behavioral models considered.

```
turtles-own [
  subject-id           ; id number of each subject
  ;; values in original lab experiment
  expPubGoodInv        ; the contribution in the original
                         experiment
  expEarnings          ; the earnings in the original experiment
  expGroupPubGood      ; the total group contribution in the
                         original experiment
  ;; variables in simulation
  pubGoodInv           ; the public good contribution of the
                         agent in the
                       ; simulation
  earnings             ; the earnings of the agent in the
                         simulation
  groupPubGood         ; the sum of group public good contribu-
                         tion in the
                       ; simulation
  ;; variables in simulation under specific agents' behavior
  ;; fixed effects
```

```
fixedEffectsGroupLag1        ; the coefficient of total group
                               investment (lag1)
                             ; estimated with the fixed
                               effects linear regression
fixedEffectsEarningsLag1     ; the coefficient of personal
                               earnings (lag1)
                             ; estimated with the fixed
                               effects linear
                             ; regression
fixedEffectsCons             ; the constant of each
                               individual, estimated with
                               fixed
                             ; effects linear regression with
                               individual dummies
;; random coefficients
rndCoeffGroupLag1            ; the coefficient of total group
                               investment (lag1)
                             ; estimated with random coeffi-
                               cients linear regression
rndCoeffEarningsLag1         ; the coefficient of personal
                               earnings (lag1)
                             ; estimated with random coeffi-
                               cients linear regression
rndCoeffCons                 ; the constant of each
                               individual, estimated with
                               random
                             ; coefficients linear regression
;; first differences
fdCoeffGroupLag1             ; the coefficient of total group
                               investment (lag1)
                             ; estimated with first
                               difference linear regression
fdCoeffEarningsLag1          ; the coefficient of personal
                               earnings (lag1) estimated
                             ; with first difference linear
                               regression
fdCoeffCons                  ; the constant of each
                               individual, estimated with
                               first
                             ; difference linear regression
fdDiffEarnings               ; the difference in earnings in
                               first difference
                             ; linear regression
fdDiffGroupPubGood           ; the difference in total group
                               contribution in first
                             ; difference linear regression
```

```
;; ordered probit
oprCoeffGroupLag1        ; the coefficient of total group
                           investment (lag1)
                         ; estimated with ordered probit
oprCoeffEarningsLag1     ; the coefficient of personal
                           earnings (lag1)
                         ; estimated with ordered probit
oprCoeffCons             ; the constant of each individual,
                           estimated with ordered
                         ; probit
oprPubGoodInvTemp        ; the temporary linear variable for
                           contribution in
                         ; ordered probit
oprDeltaEarnings         ; the earnings change (0 =
                           decrease, 1 = constant,
                         ; 2 = increase) in the ordered
                           probit
oprDeltaGroupPubGood     ; the total group contribution
                           change (0 = decrease,
                         ; 1 = constant, 2 = increase) in
                           ordered probit
;; decision trees
symPubGoodInvTemp        ; the temporary variable for
                           contribution with decision
                         ; trees
symPubGoodInvOld         ; the old value of contribution to
                           the public good
symDeltaPubGoodInv       ; the contribution change (2 =
                           decrease,
                         ; 1 = constant, 0 = increase) with
                           decision trees
symDeltaEarnings         ; the earnings change (2 =
                           decrease, 1 = constant,
                         ; 0 = increase) with decision trees
symDeltaGroupPubGood     ; the total group contribution
                           change (2 = decrease,
                         ; 1 = constant, 0 = increase) with
                           decision trees
symDeltaLag1PubGoodInv   ; the lagged contribution change
                           (2 = decrease,
                         ; 1 = constant, 0 = increase) with
                           decision
                         ; trees
symDeltaLag1Earnings     ; the lagged earnings change
                           (2 = decrease,
```

```
                              ; 1 = constant, 0 = increase)
                                with decision
                              ; trees
  symDeltaLag1GroupPubGood    ; the lagged total group
                                contribution change
                              ; (2 = decrease, 1 = constant, 0
                                = increase)
                              ; with decision trees
  ;; heuristics
  absPubGoodInvTemp           ; the temporary variable for
                                contribution with heuristics
  absDeltaEarnings            ; the change (2 = decrease, 1 =
                                constant,
                              ; 0 = increase) in earnings with
                                heuristics

  ;; reinfrocement learning
  learningCoopProb            ; the cumulative rewards of
                                behaving as altruistic
                              ; cooperator (probability)
  learningReciProb            ; the cumulative rewards of
                                behaving as reciprocator
                              ; (probability)
  learningSelfProb            ; the cumulative rewards of
                                behaving as selfish
                              ; (probability)
  learningPubGoodInvTemp      ; the temporary variable containing
                                the heuristic
                              ; explored with reinforcement
                                learning
  learningPubGoodInvTemp2     ; a second temporary variable for
                                determining
                              ; contributions with
                                reinforcement learning
  learningDeltaEarnings       ; the change (2 = decrease,
                                1 = constant,
                              ; 0 = increase) in earnings with
                                reinforcement
                              ; learning
  ;; values of the distribution of changes in contribution
  deltaMean                   ; the delta mean value of the
                                subject
  deltaStddv                  ; the delta standard deviation of
                                the subject
]
```

A.2.2   Simulation setup

The setup of a simulation is the activity by which the model is prepared and instantiated in memory. It is an activity that, obviously, must precede the actual simulation.

The first procedure is the one that is called by the "setup" button in the interface. The procedure then calls other subprocedures that take care of most of the work that has to be done to set up the model. In particular, the procedure clears the memory of previously instances of the model (eventually present); it assigns the random seed; it calls subprocedures defining global variables, agents, and interaction; and it sets the clock at the beginning of the simulation time.

```
to setup
  clear-all
  ; define random seed as the one chosen by the user
  random-seed randomSeed
  ; define initial variables and lists
  setup-globals
  ; set the shape of agents
  set-default-shape turtles "circle"
  ; setup agents
  setup-turtles
  ; setup interaction structures
  setup-interaction
  reset-ticks
end
```

Global variables are defined in the following procedure. In particular, the number of subjects is defined as in the experiment, the number of groups is computed, and the agents are assigned to the two different classrooms used in the original experiment.

```
to setup-globals
  set numSubjects 40
  set numGroups numSubjects / groupSize
  set listClass1 [1 2 3 4 5 6 7 8 9 10 11 12 13 14 15 16 17
  18 19 20]
  set listClass2 [21 22 23 24 25 26 27 28 29 30 31 32 33 34
  35 36 37 38 39 40]
  set linksList []
end
```

The creation of turtles is divided in two subprocedures, the former actually creating agents and the latter defining the color to show in the interface. The procedure ends by defining the scheme to use in the interface to show agents in space (i.e., a circle layout).

```
to setup-turtles
  setup-subjects
  setup-color
  layout-circle turtles max-pxcor - 1
end
```

The color of agent is simply set depending on the classroom where individuals were seated in the original experiment.

```
to setup-color
  ask turtles [
    ifelse subject-id < 21
    [set color green]
    [set color red]
  ]
end
```

The actual creation of agents happens in the procedure that is reported in the following where the sequence of characters {...} indicates that a part of the code has been removed to make the code more readable. Code removal regards a long list of calibrated parameters.

The procedure creates agents, it assigns calibrated values of parameters to model-level lists, and then it uses such lists to communicate individual values to agents.

```
to setup-subjects
  ; create agents and assign id
  create-turtles numSubjects
  (foreach [0 {...} 39]
   [ask turtle ? [ set subject-id ((?) + 1)]
   ])
  ; assign values of parameters for calibration to model level
    lists
  set expPubGoodInvList [0 {...} 60]
  set expEarningsList [128 {...} 55]
  set expGroupInvList [135 {...} 110]
```

```
set fixedEffectsConsList [19.84493 {…} 27.06958]
set rndCoeffGroupList [-0.019055 {…} 0.2162515]
set rndCoeffEarningsList [0.2058599 {…} 0.1361698]
set rndCoeffConsList [6.621733 {…} -7.971394]
set oprConsList [-1.197541 {…} 0]
; assign values for distribution of changes in contribution
  levels
set deltaMeanList [31.66666667 {…} 24.28571429]
set deltaStddvList [20.41241452 {…} 14.26784597]
; assign values of parameters for calibration to agents
ask turtles [
 set fixedEffectsGroupLag1 0.1750854
 set fixedEffectsEarningsLag1 -0.1754334
 set fixedEffectsCons (item (subject-id - 1)
 fixedEffectsConsList)
 set rndCoeffGroupLag1 (item (subject-id - 1)
 rndCoeffGroupList)
 set rndCoeffEarningsLag1 (item (subject-id - 1)
 rndCoeffEarningsList)
 set rndCoeffCons (item (subject-id - 1) rndCoeffConsList)
 set fdCoeffGroupLag1 -0.208851
 set fdCoeffEarningsLag1 0.4683846
 set fdCoeffCons -3.093352
 set oprCoeffGroupLag1 -0.2887042
 set oprCoeffEarningsLag1 0.4011909
 set oprCoeffCons (item (subject-id - 1) oprConsList)
 set deltaMean (item (subject-id - 1) deltaMeanList)
 set deltaStddv (item (subject-id - 1) deltaStddvList)
 ]
end
```

Similarly to the setup of agents, even the setup of the interaction structures relies on subprocedures. In particular, in this case, different procedures are called depending on the interaction structure chosen by the user of the model.

```
to setup-interaction
  ifelse interaction = "orginalGroups" [
    setup-originalGroups
  ] [
    ifelse interaction = "smallWorldStable"
    [setup-smallWorldNet]
    [setup-randomGroups]
  ]
end
```

The first subprocedure for the interaction structure simply creates a list that replicates the structure of groups in the experiment, over time. It is in fact a procedure used to replicate experimental groups in the simulation.

```
to setup-originalGroups
  ; define a list with the IDs of agents assigned
  ; to different groups in original experiment
  set linksList [4 {...} 37]
end
```

In the procedure that follows, groups are randomly drawn once or more times. In fact, if the interaction structure is random but stable, a single random group assignment is defined. If the random interaction structure is dynamic, different random groups are drawn for each simulation step.

```
to setup-randomGroups
  ; define randomly assigned groups, stable or changing
  ; every simulation step
  ifelse interaction = "randomStable"
    [
      set listClass1 shuffle listClass1
      set listClass2 shuffle listClass2
      repeat numberOfSteps [
        (foreach [0 1 2 3 4 5 6 7 8 9 10 11 12 13 14 15 16 17 18 19] [
          set linksList lput (item ? listClass1) linksList
        ])
        (foreach [0 1 2 3 4 5 6 7 8 9 10 11 12 13 14 15 16 17 18 19] [
          set linksList lput (item ? listClass2) linksList
        ])
      ]
    ]
    [
      repeat numberOfSteps [
        set listClass1 shuffle listClass1
        set listClass2 shuffle listClass2
        (foreach [0 1 2 3 4 5 6 7 8 9 10 11 12 13 14 15 16 17 18 19] [
          set linksList lput (item ? listClass1) linksList
        ])
        (foreach [0 1 2 3 4 5 6 7 8 9 10 11 12 13 14 15 16 17 18 19] [
          set linksList lput (item ? listClass2) linksList
        ])
      ]
    ]
end
```

In the case of an interaction structure in the form of a small-world network, the procedure follows the algorithm described in Section 7.2. In fact, it creates neighborhoods for each agent made by four connected neighbors. Then it rewires 30 randomly selected links over the 80 ones present in the network, by connecting one of the previous agents with a distant (i.e., not in the neighborhood) one.

The procedure ends by modifying the color of rewired links in the interface. The normal color of links is gray, and rewired links are yellow.

```
to setup-smallWorldNet
  ; create neighborhoods of 4 (connecting each node with 2
    neighbors)
  let n 0
  while [n < count turtles]
  [
    ask turtle n [create-link-with
      turtle ((n + 1) mod count turtles)
      create-link-with
      turtle ((n + 2) mod count turtles)]
    set n n + 1
  ]
  ask links [set rewired false]
  ; rewire 30 links (over the 40*2=80 existing ones)
  repeat 30 [
  let rewirable links with [ not rewired ]
    if any? rewirable [
      ask one-of rewirable [
        ; the first node remains
        let node1 end1
        ; if that node is not already connected to all others
        if [ count link-neighbors ] of end1 < (count turtles - 1)
        [
          ; pick up another random node, not a neighbor
          let node2 one-of turtles with
          [ (self != node1) and (not link-neighbor? node1) ]
          ; create new link
          ask node1 [ create-link-with node2 [
            set color yellow
            set rewired true ] ]
          ; remove old link
          die
        ]
      ]
    ]
  ]
end
```

A.2.3   Running the simulation

When the "go" button is pressed in the interface, the simulation runs. The go button is defined to repeat a procedure until any other object stops the simulation engine. The stopping of the simulation when the desired number of simulation steps is reached (i.e., when the desired simulation length is reached) is accomplished by means of the first part of the procedure. In fact, the clock of the simulation is compared to the parameter containing the number of simulation steps required by the user, and eventually, the simulation is stopped.

Then, a couple of procedures are called to make the interaction structure update, and agents are asked to make a decision on their investments. Finally, some variables are updated according to agents' decisions, and the simulation clock is moved forward.

```
to go
  ; check if the simulation should be stopped
  if ticks = numberOfSteps [
    stop
  ]
  ; update interaction structure
  update-links
  display
  ; decide public good contribution
  behave
  ; update results
  update-variables
  ; make the clock run
  tick
end
```

A.2.4   Decision-making

Agents' decision-making is governed by a procedure called "behave" that first updates data used to plot experimental results and then calls the different models of agents' behavior according to the selection made by the model user. The former activity is implemented by assigning to agents the experimental values stored in the appropriate model-level list.

The procedure ends with a check of the boundaries of public good investment. In fact, the scheme of the social dilemma modeled here specifies that public good investment should be between 0 and 60. This part of the code forces a change in investment choices in case models of behavior lead to solutions outside those boundaries.

```
to behave
  ; update choices as in experiment
  if ticks < 10 [
    ask turtles [
      set expPubGoodInv (item ((ticks * numSubjects)
        + subject-id - 1) expPubGoodInvList)
    ]
  ]
  ; update choices according to behavior
  ifelse behavior = "fixed effects" [fixedEffects] [
    ifelse behavior = "random coefficients"
    [randomCoefficients] [
      ifelse behavior = "first differences"
      [firstDifferences] [
        ifelse behavior = "ordered probit"
        [orderedProbit] [
          ifelse behavior = "decision trees"
          [decisionTrees] [
            ifelse behavior = "heuristics"
            [heuristics] [
              if behavior = "reinforcement learning"
              [reinforcementLearning]
            ]
          ]
        ]
      ]
    ]
  ]
  ; verify and force boundaries of investment choice
  ask turtles [
    if pubGoodInv > 60 [set pubGoodInv 60]
    if pubGoodInv < 0 [set pubGoodInv 0]
  ]
end
```

The first model of behavior to consider is the one calibrated with a fixed effects regression model (Section 8.3.1). The investment is set as the discrete amount of experimental currency units given by the sum between the constant and the products between coefficients and independent variables. As discussed in Chapter 8, the value of the constant is different for each agent.

This procedure and all the ones that follow and that implement behavioral models of agent decision-making (either calibrated or classified) set the choices done during the first three simulation steps equal to the ones done at the same time in the experiment by the same individuals. This choice is compulsory only for multilevel decision trees (e.g., ordered probit models would need only two initial interactions as in the experiment), but it guarantees an equal setup of the simulation under all behavioral agents for purposes of comparison.

```
to fixedEffects
    ask turtles [
    ifelse ticks < 3 [
      set pubGoodInv (item ((ticks * numSubjects)
          + subject-id - 1) expPubGoodInvList)
    ] [
      set pubGoodInv round (fixedEffectsCons
        + groupPubGood * fixedEffectsGroupLag1
        + earnings * fixedEffectsEarningsLag1)
    ]
    ]
end
```

Similarly, the procedure of the behavioral model calibrated with the random coefficients regression computes the investment using information about the independent variables of the previous interaction, which are the total contribution to the public good in the group the subject belonged to and individual earnings.

```
to randomCoefficients
    ask turtles [
    ifelse ticks < 3 [
      set pubGoodInv (item ((ticks * numSubjects)
          + subject-id - 1) expPubGoodInvList)
    ] [
      set pubGoodInv round (rndCoeffCons
        + groupPubGood * rndCoeffGroupLag1
        + earnings * rndCoeffEarningsLag1)
    ]
    ]
end
```

In the case of the procedure implementing the model calibrated with the first differences regression, the independent variables are the first differences of the variables used in the preceding behavioral models.

```
to firstDifferences
    ask turtles [
    ifelse ticks < 3 [
        set pubGoodInv (item ((ticks * numSubjects)
            + subject-id - 1) expPubGoodInvList)
    ] [
        set pubGoodInv (round (fdCoeffCons
            + fdDiffGroupPubGood * fdCoeffGroupLag1
            + fdDiffEarnings * fdCoeffEarningsLag1)) + pubGoodInv
        ]
    ]
end
```

In the case of the behavioral model calibrated with the ordered probit regression, the algorithm is slightly different. In fact, the first part similarly computes a temporary linear variable. Then, cut-points are used to decide whether contribution levels have to be decreased, kept constant, or increased. If they should be modified, the algorithm randomly draws the change in contribution level using a normal distribution calibrated with the mean and standard deviation observed for each agent in the experiment.

```
to orderedProbit
    ask turtles [
    ifelse ticks < 3 [
        set pubGoodInv (item ((ticks * numSubjects)
            + subject-id - 1) expPubGoodInvList)
    ] [
        set oprPubGoodInvTemp round (oprCoeffCons
            + oprCoeffGroupLag1 * oprDeltaGroupPubGood
            + oprCoeffEarningsLag1 * oprDeltaEarnings)
        ifelse oprPubGoodInvTemp <= -1.585069 [
            set pubGoodInv round (pubGoodInv
                - random-normal deltaMean deltaStddv)
        ] [
            ifelse oprPubGoodInvTemp <= -0.3102108 [
                set pubGoodInv pubGoodInv
            ] [
                set pubGoodInv round (pubGoodInv
                    + random-normal deltaMean deltaStddv)
            ]
        ]
    ]
    ]
end
```

The multilevel decision trees calibrated with genetic programming are modeled individually. The procedure starts by updating variables containing lagged values that are used for decision-making. Then, it defines the behavior of each agent according to the results of the behavioral analysis (as the sequence of characters {…} points out, we report here only the modeling of the first and last agents). The procedure concludes by applying changes in investment levels according to the decision chosen by behavioral algorithms and to the distribution of changes observed in the experiment.

```
to decisionTrees
    ask turtles [
      ; update specific information necessary for behavior
      if ticks > 1 [
        ifelse ticks < 3 [
          ifelse pubGoodInv < symPubGoodInvOld [
            set symPubGoodInvTemp 2
          ][
          ifelse pubGoodInv = symPubGoodInvOld [
            set symPubGoodInvTemp 1
          ][
            set symPubGoodInvTemp 0
          ]
          ]
        ][
          set symDeltaLag1PubGoodInv symDeltaPubGoodInv
        ]
        set symDeltaPubGoodInv symPubGoodInvTemp
        set symPubGoodInvOld pubGoodInv
      ]
      ; decide the action to take
      ifelse ticks < 3 [
        set pubGoodInv (item ((ticks * numSubjects)
            + subject-id - 1) expPubGoodInvList)
      ][
        ; set individual behavior for each agent
        set symPubGoodInvTemp 2
        if subject-id = 1 [
          ifelse symDeltaEarnings = 2 [
            if symDeltaEarnings != symDeltaGroupPubGood [
            set symPubGoodInvTemp 2
            ]
          ][
            set symPubGoodInvTemp symDeltaGroupPubGood
          ]
        ]
      ]
{...}
```

```
      if subject-id = 40 [
        if symDeltaGroupPubGood = 2 [
          set symPubGoodInvTemp symDeltaEarnings
        ]
      ]
      ; applying deltas
      if ticks > 2 [
        ifelse symPubGoodInvTemp = 2 [
          set pubGoodInv round (pubGoodInv
            - random-normal deltaMean deltaStddv)
        ][
          ifelse symPubGoodInvTemp = 1 [
            set pubGoodInv pubGoodInv
          ][
            set pubGoodInv round (pubGoodInv
              + random-normal deltaMean deltaStddv)
          ]
        ]
      ]
    ]
  ]
end
```

The procedure implementing the results of the classification over a limited set of heuristics firstly selects agents, and it then applies the simple algorithms of heuristics.

```
to heuristics
    ask turtles [
    ifelse ticks < 3 [
      set pubGoodInv (item ((ticks * numSubjects)
          + subject-id - 1) expPubGoodInvList)
    ][
      ; constant players (selfish and altruistic players,
      ; depending upon their starting values)
      if subject-id = 2 or subject-id = 4 or subject-id = 17
      or subject-id = 18 or subject-id = 29 or subject-id = 33 [
        set absPubGoodInvTemp 1
      ]
      ; selfish players
      if subject-id = 3 or subject-id = 10 or subject-id = 18
      or subject-id = 25 or subject-id = 31 or subject-id = 35
      or subject-id = 39 [
        set absPubGoodInvTemp 2
      ]
```

```
; altruistic players
if subject-id = 13 or subject-id = 26 or subject-id = 40 [
  set absPubGoodInvTemp 0
]
; reciprocating players
if subject-id = 1 or subject-id = 5 or subject-id = 6
or subject-id = 7 or subject-id = 8 or subject-id = 11
or subject-id = 12 or subject-id = 14 or subject-id = 15
or subject-id = 19 or subject-id = 20 or subject-id = 21
or subject-id = 23 or subject-id = 24 or subject-id = 27
or subject-id = 29 or subject-id = 30 or subject-id = 32
or subject-id = 34 or subject-id = 36 or subject-id = 38 [
  set absPubGoodInvTemp absDeltaEarnings
]
; unclassified, random players
if subject-id = 9 or subject-id = 22 or subject-id = 37 [
  set absPubGoodInvTemp random 3
]
; applying deltas
if ticks > 2 [
  ifelse absPubGoodInvTemp = 2 [
    set pubGoodInv round (pubGoodInv
      - random-normal deltaMean deltaStddv)
  ][
    ifelse absPubGoodInvTemp = 1 [
      set pubGoodInv pubGoodInv
    ][
      set pubGoodInv round (pubGoodInv
        + random-normal deltaMean deltaStddv)
    ]
  ]
]
    ]
  ]
]
end
```

The procedure implementing reinforcement learning is different from the preceding ones since it is aimed at modeling a more sophisticated decision-making in agents.

In particular, it starts by randomly selecting a behavior in the first three steps of the simulation. The aim of this part is to allow some exploration of the three possible kinds of behavior considered here. Further, at the end of the procedure, experimentally observed deltas are applied except for the first simulation step where a random discrete value of contribution level is chosen in the boundary of feasibility for that variable.

After the first three simulation steps, the procedure implements the choice according to the rewards received by the three different strategies in preceding simulation steps. Cumulative rewards are normalized and used as probabilities. In particular, a random number between 0 and 1 is drawn, and using probabilities as in a random wheel, the kind of behavior to adopt is selected.

```
to reinforcementLearning
  ask turtles [
    ifelse ticks < 3 [
      set learningPubGoodInvTemp random 3
    ][
      ; for normalizing probabilities
      let tempSum (learningCoopProb + learningReciProb
        + learningSelfProb)
      ; chosing abstract strategy (0 altruistic, 1
        reciprocating, 2 selfish)
      let temp random-float 1
      ifelse temp < learningCoopProb / tempSum [
        set learningPubGoodInvTemp 0
      ] [
        ifelse temp <
        (learningCoopProb + learningReciProb)  / tempSum [
          set learningPubGoodInvTemp 1
        ] [
          set learningPubGoodInvTemp 2
        ]
      ]
      ; modify strategy for reciprocating players
      ; in order to reciprocate group behavior
      ifelse learningPubGoodInvTemp = 1 [
        set learningPubGoodInvTemp2 learningDeltaEarnings
      ] [
        set learningPubGoodInvTemp2 learningPubGoodInvTemp
      ]
    ]
    ; applying deltas, but choosing random contribution level
    ; at first step
    ifelse ticks > 0 [
        ifelse learningPubGoodInvTemp2 = 2 [
          set pubGoodInv round (pubGoodInv
            - random-normal deltaMean deltaStddv)
        ][
          ifelse learningPubGoodInvTemp2 = 1 [
            set pubGoodInv learningDeltaEarnings
          ][
```

```
                set pubGoodInv round (pubGoodInv
                   + random-normal deltaMean deltaStddv)
                ]
             ]
          ]
          [
             set pubGoodInv random 61
          ]
       ]
    end
```

A.2.5   Updating interaction structure and other variables

At the beginning of each simulation step, the interaction structure is updated if needed (see the "go" procedure). In particular, the interaction structure is not updated if the user has chosen a small-world stable social network, a random but stable group allocation, or if the choice has been to use the original groups and the simulation step is beyond the rounds observed in the experiment. In this latter case, the interaction structure remains "frozen" as observed in the last round of the experiment.

When links between agents are updated, the algorithm uses the information stored in model-level lists defined at the moment of the setup of the simulation.

```
to update-links
   ifelse ((interaction = "originalGroups"
         and ticks >= 10)
   or interaction = "smallWorldStable"
   or (interaction = "randomStable"
      and ticks > 0)) [
   ] [
      clear-links
      set groupCounter 0
      set tickConstant (ticks * numSubjects)
      while [groupCounter < numGroups] [
         set subjectCounter 0
         while [subjectCounter < (groupSize - 1)] [
            set positionFrom (tickConstant
               + (groupCounter * groupSize) + subjectCounter)
            set positionTo (tickConstant
               + (groupCounter * groupSize) + groupSize - 1)
            while [positionFrom < positionTo] [
```

```
            ask turtle (
              (item positionFrom linksList) - 1) [
              create-link-with
              turtle ((item positionTo linksList) - 1)
            ]
            set positionTo (positionTo - 1)
          ]
        set subjectCounter (subjectCounter + 1)
      ]
      set groupCounter (groupCounter + 1)
    ]
  ]
end
```

The procedure that follows manages the update of the two other relevant
variables of the game of voluntary provision of public goods, that is to say the
total contribution to the public good at the group level and the computation of
individual earnings. Finally, the procedure asks to update cumulative rewards
if reinforcement learning is used in the simulation.

```
to update-variables
  update-groupInv
  update-earnings
  if behavior = "reinforcement learning" [
    update-learning
  ]
end
```

Because all agents belong to a social network even if fully connected
groups are used as interaction structure, the computation of the sum of contributions to the public good is straightforward and presented at the end of the
procedure. Each agent asks his neighbors, which are the agents directly linked
to him, to communicate their contribution levels, and he sums them with its
own contribution. When fully connected groups are used, this amount will be
the same for each member of the same group. When using a small-world network, the mechanism is slightly different, but it still relies on the network
structure and on the connection with neighbors. In fact, with a small-world
network, groups are overlapped and defined by the network structure. Further,
they are not all of the same size.

At the beginning of the procedure, there is the selection of experimental
data, and then specific variables associated with total group investment are
updated for different behavioral models.

```
to update-groupInv
  ; for experiment data
  if ticks < 10 [
    ask turtles [
      set expGroupPubGood (item ((ticks * numSubjects)
          + subject-id - 1) expGroupInvList)
    ]
  ]
  ; update information used by specific
  ; agents' behavioral models
  ifelse behavior = "first differences" [
    ask turtles [
      if ticks != 0 [
        set fdDiffGroupPubGood (pubGoodInv
          + sum [pubGoodInv] of link-neighbors) - groupPubGood
      ]
    ]
  ]
  [
    ifelse behavior = "ordered probit" [
      ask turtles [
        if ticks != 0 [
          ifelse (pubGoodInv
            + sum [pubGoodInv] of link-neighbors)
            < groupPubGood [
            set oprDeltaGroupPubGood 0
          ][
          ifelse (pubGoodInv
            + sum [pubGoodInv] of link-neighbors)
            = groupPubGood [
            set oprDeltaGroupPubGood 1
          ][
          set oprDeltaGroupPubGood 2
          ]
          ]
        ]
      ]
    ]
    [
      if behavior = "decision trees" [
        ask turtles [
          if ticks != 0 [
            if ticks != 1 [
              set symDeltaLag1GroupPubGood symDeltaGroupPubGood
            ]
            ifelse (pubGoodInv
```

```
                + sum [pubGoodInv] of link-neighbors)
                < groupPubGood [
                set symDeltaGroupPubGood 2
              ][
              ifelse (pubGoodInv
                + sum [pubGoodInv] of link-neighbors)
                = groupPubGood [
                set symDeltaGroupPubGood 1
              ][
              set symDeltaGroupPubGood 0
              ]
              ]
            ]
          ]
        ]
      ]
    ]
    ; update sum of group contributions for all behavioral
      models
    ask turtles [
      set groupPubGood (pubGoodInv + sum [pubGoodInv] of
      link-neighbors)
    ]
end
```

The computation of earnings is organized as the computation of total public good investment in the preceding procedure. Earnings, in particular, are computed at the end of the procedure using returns for the private investment and the public one.

```
to update-earnings
  ; for experiment data
  if ticks < 10 [
    ask turtles [
      set expEarnings (item ((ticks * numSubjects)
        + subject-id - 1) expEarningsList)
    ]
  ]
  ; update information used by specific agents' behavioral
    models
  ifelse behavior = "first differences" [
    ask turtles [
      if ticks != 0 [
        set fdDiffEarnings (round ((groupPubGood
```

```
                  * publicGoodReturn * 100) +
          ((60 - pubGoodInv) * privateGoodReturn
            * 100))) - earnings
      ]
    ]
  ]
  [
    ifelse behavior = "ordered probit" [
      ask turtles [
        if ticks != 0 [
          ifelse round ((groupPubGood
              * publicGoodReturn * 100) +
          ((60 - pubGoodInv) * privateGoodReturn
            * 100)) < earnings [
            set oprDeltaEarnings 0
          ][
          ifelse round ((groupPubGood
              * publicGoodReturn * 100) +
          ((60 - pubGoodInv) * privateGoodReturn
            * 100)) = earnings [
            set oprDeltaEarnings 1
          ][
          set oprDeltaEarnings 2
          ]
          ]
        ]
      ]
    ]
    [
      ifelse behavior = "decision trees" [
        ask turtles [
          if ticks != 0 [
            if ticks != 1 [
              set symDeltaLag1Earnings symDeltaEarnings
            ]
            ifelse round ((groupPubGood
                * publicGoodReturn * 100) +
            ((60 - pubGoodInv) * privateGoodReturn * 100))
            < earnings [
              set symDeltaEarnings 2
            ][
            ifelse round ((groupPubGood
                * publicGoodReturn * 100) +
            ((60 - pubGoodInv) * privateGoodReturn * 100))
            = earnings [
              set symDeltaEarnings 1
```

```
        ] [
        set symDeltaEarnings 0
        ]
        ]
      ]
    ]
  ]
  [
    ifelse behavior = "heuristics" [
      ask turtles [
        if ticks != 0 [
          ifelse round ((groupPubGood
              * publicGoodReturn * 100) +
          ((60 - pubGoodInv) * privateGoodReturn
          * 100)) < earnings [
            set absDeltaEarnings 2
          ] [
          ifelse round ((groupPubGood
              * publicGoodReturn * 100) +
          ((60 - pubGoodInv) * privateGoodReturn
          * 100)) = earnings [
            set absDeltaEarnings 1
          ] [
          set absDeltaEarnings 0
          ]
          ]
        ]
      ]
    ]
    [
      if behavior = "reinforcement learning" [
        ask turtles [
          if ticks != 0 [
            ifelse round ((groupPubGood
                * publicGoodReturn * 100) +
            ((60 - pubGoodInv) * privateGoodReturn
            * 100)) < earnings [
              set learningDeltaEarnings 2
            ] [
            ifelse round ((groupPubGood
                * publicGoodReturn * 100) +
            ((60 - pubGoodInv) * privateGoodReturn
            * 100)) = earnings [
              set learningDeltaEarnings 1
            ] [
```

```
                set learningDeltaEarnings 0
                ]
                ]
            ]
          ]
        ]
       ]
      ]
     ]
    ]
   ; update earnings for all behavioral models
   ask turtles [
     set earnings round ((groupPubGood * publicGoodReturn * 100)
        + ((60 - pubGoodInv) * privateGoodReturn * 100))
   ]
 end
```

The last procedure, concerning reinforcement learning, takes care of updating the cumulative rewards associated with the different kinds of behavior available, and those are used as probabilities of selection. In summary, earnings are added to the cumulative value associated with the action that has been explored.

```
to update-learning
  ask turtles [
    if ticks != 0 [
      ; add earnings to probability values
      ; (i.e., cumulated earnings by strategy)
      ifelse learningPubGoodInvTemp = 0 [
        set learningCoopProb (learningCoopProb + earnings)
      ][
        ifelse learningPubGoodInvTemp = 1 [
          set learningReciProb (learningReciProb + earnings)
        ][
          set learningSelfProb (learningSelfProb + earnings)
        ]
      ]
    ]
  ]
end
```

References

Abrahamson D. and Wilensky U. (2005), Piaget? Vygotsky? I'm game! Agent-based modeling for psychology research, Paper presented at the annual meeting of the Jean Piaget Society, Vancouver, Canada, June 2005.

Abrams M. (2013), A moderate role for cognitive models in agent-based modeling of cultural change, *Complex Adaptive Systems Modeling*, 1: 16.

Acock A.C. (2013), *Discovering Structural Equation Modeling Using Stata*, College Station, TX: Stata Press.

Alam S.J., Geller A., Meyer R., Werth B. (2010), Modeling contextualized reasoning in complex societies with "endorsements", *Journal of Artificial Societies and Social Simulation*, 13(4): 6.

Alchian A.A. (1950), Uncertainty, evolution and economic theory, *Journal of Political Economy*, 58: 211–221.

Anderson P.W. (1972), More is different, *Science*, 177(4047): 393–396.

Anderson J.R., Fincham J.M., Qin Y., Stocco A. (2008), A central circuit of the mind, *Trends in Cognitive Science*, 12(4): 136–143.

Andreoni J. (1988), Why free ride? Strategies and learning in public goods experiments, *Journal of Public Economics*, 37(3): 291–304.

Andreoni J. (1990), Impure altruism and donations to public goods: a theory of warm-glow giving?, *Economic Journal*, 100(401): 464–477.

Andreoni J. (1995a), Warm-glow vs. cold prickle: the effect of positive and negative framing on cooperation in experiments, *Quarterly Journal of Economics*, 110(1): 1–21.

Andreoni J. (1995b), Cooperation in public-goods experiments: kindness or confusion?, *The American Economic Review*, 85(4): 891–904.

Andreoni J. and Croson R. (2008), Partners versus strangers: random rematching in public goods experiments, in Plott C.R. and Smith V.L. (eds), *Handbook of Experimental Economics Results*, Amsterdam: North-Holland, 776–783.

Behavioral Computational Social Science, First Edition. Riccardo Boero.
© 2015 John Wiley & Sons, Ltd. Published 2015 by John Wiley & Sons, Ltd.

Arthur W.B. (1990), A learning algorithm that replicates human learning, Santa Fe Institute Working Paper 90–026.

Arthur W.B. (1994), Inductive reasoning and bounded rationality, *The American Economic Review*, 84(2): 406–411.

Atkinson R. and Flint J. (2001), Accessing hidden and hard-to-reach populations: snowball research strategies, Social Research Update, Issue 33, Department of Sociology, University of Surrey, Surrey.

Axelrod R. (1984), *The Evolution of Cooperation*, New York: Basic Books.

Axelrod R. (1997a), *The Complexity of Cooperation: Agent-Based Models of Competition and Collaboration*, Princeton, NJ: Princeton University Press.

Axelrod R. (1997b), Advancing the art of simulation in the social sciences, *Complexity*, 3(2): 16–22.

Axelrod R. and Tesfatsion L. (2005), A guide for newcomers to agent-based modeling in the social sciences, in Judd K.L. and Tesfatsion L. (eds), *Handbook of Computational Economics, Vol. 2: Agent-Based Computational Economics*, Amsterdam: North-Holland, 1647–1658.

Axtell R. (2000), Why agents? On the varied motivations for agent computing in the social sciences, in Macal C.M. and Sallach D. (eds), *Proceedings of the Workshop on Agent Simulation: Applications, Models and Tools*, Chicago, IL: Argonne National Laboratory, 3–24.

Bagnoli M. and McKee M. (1991), Voluntary contribution games: efficient private provision of public goods, *Economic Inquiry*, 29: 351–366.

Bainbridge W.S. (2007), The scientific research potential of virtual worlds, *Science*, 317: 472–476.

Baltagi B.H. (2013), *Econometric Analysis of Panel Data*, 5th edition, Chichester: John Wiley & Sons, Ltd.

Banks J., Carson J.S., Nelson B.L., Nicol D.M. (2000), *Discrete-Event System Simulation*, New York: Prentice Hall.

Barabási A.L. and Albert R. (1999), Emergence of scaling in random networks, *Science*, 286(5439): 509–512.

Barreteau O. and Abrami G. (2007), Variable time scales, agent-based models, and role-playing games: the PIEPLUE river basin management game, *Simulation & Gaming*, 38(3): 364–381.

Bedau M.A. (1998), Philosophical content and method of artificial life, in Bynum T.W. and Moor J.H. (eds), *The Digital Phoenix: How Computers are Changing Philosophy*, Oxford: Basil Blackwell.

Bedau M.A. (1999), Can unrealistic computer models illuminate theoretical biology?, in Wu A.S. (ed.), *Proceedings of the 1999 Genetic and Evolutionary Computation Conference*, San Francisco, CA: Morgan Kaufmann.

Bender E.A. and Canfield E.R. (1978), The asymptotic number of labeled graphs with given degree sequences, *Journal of Combinatorial Theory A*, 24: 296–307.

Boero R. (2002), SWIEE—a Swarm web interface for experimental economics, in Luna F. and Perrone A. (eds), *Agent Based Methods in Economic and Finance: Simulations in Swarm*, Boston, MA: Kluwer Academic Publishers.

Boero R. (2007), The social mechanism of public good provision: analytically researching social dilemmas with empirically founded agent based models, PhD thesis, University of Surrey.

Boero, R. (2011), Food quality as a public good: cooperation dynamics and economic development in a rural community. *Mind & Society*, 10(2): 203–215.

Boero R. and Novarese M. (2012), Feedback and learning, in Seel N.M. (ed.), *Encyclopedia of the Sciences of Learning*, New York: Springer, 1282–1285.

Boero R. and Squazzoni F. (2005), Does empirical embeddedness matter? Methodological issues on agent-based models for analytical social science, *Journal of Artificial Societies and Social Simulation*, 8(4): 6.

Boero R., Castellani M., Squazzoni F. (2004), Micro behavioural attitudes and macro technological adaptation in industrial districts: an agent-based prototype, *Journal of Artificial Societies and Social Simulation*, 7(2).

Boero R., Castellani M., Squazzoni F. (2008), Individual behavior and macro social properties. An agent-based model, *Computational and Mathematical Organization Theory*, 14(2): 156–174.

Boero R., Bravo G., Castellani M., Squazzoni F. (2010), Why bother with what others tell you? an experimental data-driven agent-based model, *Journal of Artificial Societies and Social Simulation*, 13(3): 6.

Bosse T., Gerritsen C., Treur J. (2009), Towards integration of biological, psychological and social aspects in agent-based simulation of violent offenders, *Simulation*, 85(10): 635–660.

Boudon R. (1998), Social mechanisms without black boxes, in Hedström P. and Swedberg R. (eds), *Social Mechanisms: An Analytical Approach to Social Theory*, Cambridge: Cambridge University Press, 172–203.

Bratman M.E. (1987), *Intentions, Plans, and Practical Reason*, Cambridge, MA: Harvard University Press.

Bravo G., Squazzoni F., Boero R. (2012), Trust and partner selection in social networks: an experimentally grounded model, *Social Networks*, 34(4): 481–492.

Brown-Kruse J. and Hummels D. (1993), Gender effects in laboratory public goods contribution: do individuals put their money where their mouth is?, *Journal of Economic Behavior and Organization*, 22: 255–267.

Camerer C. (2006), Behavioral economics, in Blundell R., Newey W., Persson T. (eds), *Advances in Economics and Econometrics*, Cambridge: Cambridge University Press, 181–214.

Camerer C. (2008), The potential of neuroeconomics, *Economics and Philosophy*, 24: 369–379.

Camerer C., Loewenstein G., Prelec D. (2005), Neuroeconomics: how neuroscience can inform economics, *Journal of Economic Literature*, 43: 9–64.

Cameron A.C. and Trivedi P.K. (2013), *Regression Analysis of Count Data*, 2nd edition, Cambridge: Cambridge University Press.

Castelfranchi C. (1998), Simulating with cognitive agents: the importance of cognitive emergence, in Sichman J.S., Conte R., Gilbert N. (eds), *Multi-Agent Systems and Agent-Based Simulation*, Lecture Notes in Computer Science 1534, Berlin Heidelberg: Springer, 26–44.

Chamberlin J. (1974), Provision of collective goods as a function of group size, *American Political Science Review*, 68: 707–716.

Chiang Y. (2010), Self-interested partner selection can lead to the emergence of fairness, *Evolution and Human Behavior*, 31: 265–270.

Ciriolo E. (2011), Behavioural economics in the European Commission: past, present and future, *Oxera Agenda*, January: 1–5.

Coleman J.S. (1986), Social theory, social research, and a theory of action, *American Journal of Sociology*, 91: 1309–1335.

Coleman J.S., Katz E., Menzel H. (1957), The diffusion of an innovation among physicians, *Sociometry*, XX: 253–270.

Coleman J.S., Katz E., Menzel H. (1966), *Medical Innovation*, Indianapolis, IN: Bobbs-Merril.

Conte R. (2002), Agent-based modeling for understanding social intelligence, *PNAS*, 99(3): 7189–7190.

Conte R. and Castelfranchi C. (1995), *Cognitive and Social Action*, London: University College of London Press.

Conte R. and Paolucci M. (2001), Intelligent social learning, *Journal of Artificial Societies and Social Simulation*, 4(1): 3.

Conte R. and Paolucci M. (2002), *Reputation in Artificial Societies: Social Beliefs for Social Order*, Dordrecht: Kluwer Academic.

Conte R. and Paolucci M. (2014), On agent-based modeling and computational social science, *Frontiers in Psychology*, 4: 668.

Conte R., Andrighetto G., Campenni M. (2013), *Minding Norms: Mechanisms and Dynamics of Social Order in Agent Societies*, Oxford: Oxford University Press.

Corten R. (2011), Visualization of social networks in Stata using multidimensional scaling, *The Stata Journal*, 11(1): 52–63.

Corten R. (2014), *Computational Approaches to Studying the Co-evolution of Networks and Behavior in Social Dilemmas*, Chichester: John Wiley & Sons, Ltd.

Costa Pereira C., Mauri A., Tettamanzi G.B. (2009), Cognitive-agent-based modeling of a financial market, *Proceedings of the 2009 IEEE/WIC/ACM International Joint Conference on Web Intelligence and Intelligent Agent Technology*, Washington, DC: IEEE Computer Society, 20–27.

Davis J.B. (2004), The agency-structure model and the embedded individual in heterodox economics, in Lewis P. (ed.), *Transforming Economics, Perspective on the Critical Realist Project*, New York: Routledge, 132–151.

Davis J.B. (2010), Neuroeconomics: constructing identity, *Journal of Economic Behavior & Organization*, 76(3): 574–583.

De Leeuw J.R. (2015), jsPsych: a JavaScript library for creating behavioral experiments in a web browser, *Behavior Research Methods*, 47(1): 1–12.

De Leeuw E.D., Hox J., Dillman D. (2008), *International Handbook of Survey Methodology*, European Association of Methodology Series, London: Routledge.

Di Paolo E.A., Noble J., and Bullock S. (2000), Simulation models as opaque thought experiments, *Artificial Life VII: the Seventh International Conference on the Simulation and Synthesis of Living Systems*, August 1–6, Reed College, Portland, OR.

Dignum V., Tranier J, Dignum F. (2010), Simulation of intermediation using rich cognitive agents, *Simulation Modelling Practice and Theory*, 18(10): 1526–1536.

Dolan P., Halpern D., King D., Vlaev I. (2010), Mindspace: influencing behaviour through public policy, Discussion document, London: Institute for Government, UK Cabinet Office.

Dubé L., Bechara A., Böckenholt U., et al. (2008), Towards a brain-to-society systems model of individual choice, *Marketing Letters*, 19(3–4): 323–336.

Duffy J. (2001), Learning to speculate: experiments with artificial and real agents, *Journal of Economic Dynamics & Control*, 25: 295–319.

Duffy J. (2006), Agent-based models and human subject experiments, in Testfatsion L. and Judd K.L. (eds), *Handbook of Computational Economics, Volume 2*, Amsterdam/New York: Elsevier, 949–1011.

Ebenhoh E. and Pahl-Wostl C. (2008), Agent behavior between maximization and cooperation, *Rationality and Society*, 20(2): 227–252.

Elias N. (1991 [1939]), *The Society of Individuals*, Oxford: Blackwell.

Elster J. (1983), *Explaining Technical Change*, Cambridge: Cambridge University Press.

Elster J. (1998), A plea for mechanisms, in Hedström P. and Swedberg R. (eds), *Social Mechanisms: An Analytical Approach to Social Theory*, Cambridge: Cambridge University Press, 45–73.

Engler J. and Kusiak A. (2011), Modeling an innovation ecosystem with adaptive agents, *International Journal of Innovation Science*, 3(2): 55–67.

Epstein J.M. (1999), Agent-based computational models and generative social science, *Complexity*, 4(5): 41–60.

Epstein J.M. and Axtell R. (1994), Agent-based modeling: understanding our creations, *The Bulletin of the Santa Fe Institute*, 9(4): 28–32.

Epstein J.M. and Axtell R. (1996), *Growing Artificial Societies—Social Science from the Bottom Up*, Cambridge, MA: MIT Press.

Erdös P. and Rényi A. (1959), On random graphs, *Publicationes Mathematicae Debrecen*, 6: 290–297.

Erev I. and Roth A.E. (1998), Predicting how people play games: reinforcement learning in experimental games with unique, mixed strategy equilibria, *The American Economic Review*, 88(4): 848–881.

Erickson B.H. and Nosanchuk T.A. (1983), Applied network sampling, *Social Networks*, 5: 367–382.

Evans T., Sun W., Kelley H. (2006), Spatially explicit experiments for the exploration of land-use decision making dynamics, *International Journal of Geographical Information Science*, 20(9): 1013–1037.

Eve R.A., Horsfall S., Lee M.E. (eds) (1997), *Chaos, Complexity and Sociology: Myths, Models and Theories*, London: Sage Publications.

Falk A. and Heckman J.J. (2009), Lab experiments are a major source of knowledge in the social sciences, *Science*, 326(5952): 535–538.

Farmer J.D., Shubik M., Smith E. (2005), Economics: the next physical science? Cowles Foundation Discussion Paper no. 1520, June 2005.

Ferber J., Stratulat T., Tranier J. (2009), Towards an integral approach of organizations: the MASQ approach, in Dignum V. (ed.), *Multi-Agent Systems: Semantics and Dynamics of Organizational Models*, Hershey, PA: IGI.

Fischbacher U. (2007), z-Tree: Zurich toolbox for ready-made economic experiments, *Experimental Economics*, 10(2): 171–178.

Fleetwood S. (ed.) (1999), *Critical Realism in Economics: Development and Debate*, London: Routledge.

Forrester J.W. (1961), *Industrial Dynamics*, Cambridge, MA: Productivity Press.

Forrester J.W. (1968), *Principles of Systems*, Cambridge, MA: Productivity Press.

Foster J. (2004), Why is economics not a complex systems science? Discussion Paper no. 336, School of Economics, University of Queensland, Australia.

Friston K.J. and Dolan R.J. (2009), Computational and dynamic models in neuroimaging, *Neuroimage*, 52(3): 752–765.

Gigerenzer G. and Selten R. (eds) (2001), *Bounded Rationality: The Adaptive Toolbox*, Cambridge, MA: The MIT Press.

Gilbert E.N. (1959), Random graphs, *Annals of Mathematical Statistics*, 30(4): 1141–1144.

Gilbert N. (2002), Varieties of emergence, in Sallach D. (ed.), *Social Agents: Ecology, Exchange, and Evolution. Agent 2002 Conference*. University of Chicago and Argonne National Laboratory, 41–54.

Gilbert N. and Terna P. (2000), How to build and use agent-based models in social science, *Mind & Society* 1: 57–72.

Gilbert N. and K.G. Troitzsch (1999), *Simulation for the Social Scientist*, 2nd edition (2005), Buckingham: Open University Press.

Glimcher P.W. (2003), *Decisions, Uncertainty, and the Brain: The Science of Neuroeconomics*, Cambridge, MA: The MIT Press.

Goldstone R.L. and Janssen M.A. (2005), Computational models of collective behavior, *Trends in Cognitive Sciences*, 9(9): 424–430.

Goldstone R.L., Roberts M.E., Gureckis T.M. (2008), Emergent processes in group behavior, *Current Directions in Psychological Science*, 17(1): 10–15.

Granovetter M. (1978), Threshold models of diffusion and collective behaviour, *Journal of Mathematical Sociology*, 83: 1420–1443.

Grimm V. and Mengel F. (2009), Cooperation in viscous populations: experimental evidence, *Games and Economic Behavior*, 66: 202–220.

Grimm V., Berger U., Bastiansen F., et al. (2006), A standard protocol for describing individual-based and agent-based models, *Ecological Modelling*, 198: 115–126.

Grimm V., Berger U., DeAngelis D.L., Polhill J.G., Giske J., Railsback S.F. (2010), The ODD protocol: a review and first update, *Ecological Modelling*, 221: 2760–2768.

Hailu A. and Schilizzi S. (2004), Are auctions more efficient than fixed price schemes when bidders learn?, *Australian Journal of Management*, 29(2): 147–168.

Hargreaves H.S. (2004), Critical realism and the heterodox tradition in economics, in Lewis P. (ed.), *Transforming Economics, Perspective on the Critical Realist Project*, New York: Routledge, 152–166.

Harré R. (1970), *The Principles of Scientific Thinking*, Chicago, IL: University of Chicago Press.

Hayashi F. (2000), *Econometrics*, Princeton, NJ: Princeton University Press.

Hayek F.A. (1945), The use of knowledge in society, *American Economic Review*, 35(4): 519–530.

Hayek F.A. (1952), *The Sensory Order: An Inquiry into the Foundations of Theoretical Psychology*, Chicago, IL: The University of Chicago Press.

Heckbert S. (2009), Experimental economics and agent-based models, *Proceedings of the 18th World IMACS Congress*, July 13–17, 2009, Cairns, Australia.

Heckbert S., Baynes T., Reeson A. (2010), Agent-based modeling in ecological economics, *Annals of the New York Academy of Sciences*, 1185: 39–53.

Hedström P. (1998), Rational imitation, in Hedström P. and Swedberg R. (eds), *Social Mechanisms: An Analytical Approach to Social Theory*, Cambridge: Cambridge University Press, 306–327.

Hedström P. and Swedberg R. (eds) (1998a), *Social Mechanisms: An Analytical Approach to Social Theory*, Cambridge: Cambridge University Press.

Hedström P. and Swedberg R. (1998b), Social mechanisms: an introductory essay, in Hedström P. and Swedberg R. (eds), *Social Mechanisms: An Analytical Approach to Social Theory*, Cambridge: Cambridge University Press, 1–31.

Hendriks A. (2012), SoPHIE—Software Platform for Human Interaction Experiments, University of Osnabrück Working Paper.

Hernes G. (1998), Real virtuality, in Hedström P. and Swedberg R. (eds), *Social Mechanisms: An Analytical Approach to Social Theory*, Cambridge: Cambridge University Press, 74–101.

Holland J.H. (1975), *Adaptation in Natural and Artificial Systems*, Ann Arbor, MI: University of Michigan Press.

Holland J.H. (1992), *Adaptation in Natural and Artificial Systems*, 2nd edition, Cambridge, MA: The MIT Press.

Holland J.H. (2006), Studying complex adaptive systems, *Journal of System Science and Complexity*, 19(1): 1–8.

Holland J.H. and Miller J.H. (1991), Artificial adaptive agents in economic theory, *The American Economic Review*, 81(2): 365–371.

Holland J.H., Holyoak K.J., Nisbett R.E., Thagard P. (1986), *Induction: Processes of Inference, Learning and Discovery*, Cambridge: MIT Press.

Isaac R.M. and Walker J.M. (1988), Group size effects in public goods provision: the voluntary contributions mechanism, *Quarterly Journal of Economics*, 103(1): 179–200.

Isaac R.M., Walker J.M., Arlington A.W. (1994), Group size effects in public goods provision: experimental evidence utilizing large groups, *Journal of Public Economics*, 51(4): 595–614.

Jackson M.O. (2008), *Social and Economic Networks*, Princeton, NJ: Princeton University Press.

Jager W. and Janssen M.A. (2012), An updated conceptual framework for integrated modeling of human decision making: The Consumat II, Paper presented at the Workshop Complexity in the Real World @ ECCS 2012—from policy intelligence to intelligent policy, Brussels, 5–6 September.

Janssen M.A. and Ostrom E. (2006), Empirically based, agent-based models, *Ecology and Society*, 11(2): 37.

Janssen M.A., Radtke N.P., Lee A. (2009), Pattern-oriented modeling of commons dilemma experiments, *Adaptive Behavior*, 17(6): 508–523.

Joos P., Vanhoof K., Ooghe H., Sierens N. (1998), Credit classification: a comparison of logit models and decision trees, *Proceedings Notes of the Workshop on Application of Machine Learning and Data Mining in Finance, 10th European Conference on Machine Learning*, April 24, 1998, Chemnitz, Germany, 59–72.

Junges R. and Klügl F. (2013), Learning tools for agent-based modeling and simulation, *Künstl Intell*, 27: 273–280.

Kemeny J.G. and Snell J.L. (1960), *Finite Markov Chains*, vol. 356, Princeton, NJ: van Nostrand.

Kennedy W.G. and Bugajska M. (2010), Integrating fast and slow cognitive processes, in Salvucci D.D. and Gunzelmann G. (eds), *Proceedings of the 10th International Conference on Cognitive Modeling*, Philadelphia, PA: Drexel University, 121–126.

Killgore W.D.S., Young A.D., Femia L.A., Bogorodzki P., Rogowska J., Yurgelun-Todd D.A. (2003), Cortical and limbic activation during viewing of high- versus low-calorie foods, *Neuroimage*, 19(4): 1381–1394.

Kim O. and Walker M. (1984), The free rider problem: experimental evidence, *Public Choice*, 43: 3–24.

Kirk R.E. (2013), *Experimental Design: Procedures for the Behavioral Sciences*, 4th edition, Thousand Oaks, CA: Sage Publications Inc.

Kirman A. (2010), Learning in agent based models, Document de Travail 2010-57 GREQAM, Universités d'Aix-Mairselle II et III.

Kirman A. and Vriend N.J. (2001), Evolving market structure: an ACE model of price dispersion and loyalty, *Journal of Economic Dynamics and Control*, 25(3–4): 459–502.

Koza J.R. (1992), *Genetic Programming: On the Programming of Computers by Means of Natural Selection*, Cambridge, MA: The MIT Press.

Kuhn T.S. (1977), *A Function for Thought Experiments, in The Essential Tension: Selected Studies in Scientific Tradition and Change*, Chicago, IL: University of Chicago Press.

Latané B. and Nowak A. (1997), Self-organizing social systems: necessary and sufficient conditions for the emergence of clustering, consolidation and continuing diversity, in G.A. Barnett and F.J. Boster, eds., *Progress in Communication Sciences*, Norwood, NJ: Ablex Publishing Company.

Lave C.A. and March J.G. (1975), *An Introduction to Models in the Social Sciences*. New York: Harper & Row.

Ledyard J.O. (1995), Public goods: a survey of experimental research, in Kagel J.H. and Roth A.E. (eds), *The Handbook of Experimental Economics*, Princeton, NJ: Princeton University Press.

Luke S., Cioffi-Revilla C., Panait L., Sullivan K., Balan G. (2005), MASON: a multi-agent simulation environment, *Simulation: Transactions of the society for Modeling and Simulation International*, 82(7): 517–527.

Marewski J.N. and Schooler L.J. (2011), Cognitive niches: an ecological model of strategy selection, *Psychological Review*, 118(3): 393–437.

Marwell G. and Ames R. (1979), Experiments on the provision of public goods I: resources, interest, group size, and the free rider problem, *American Journal of Sociology*, 84(6): 1335–1360.

Mathôt S., Schreij D., Theeuwes J. (2012), OpenSesame: an open-source, graphical experiment builder for the social sciences, *Behavior Research Methods*, 44(2): 314–324.

Merton R.K. (1949), *Social Theory and Social Structure*, New York: Free Press.

Milgram S. (1977), *The Individual in a Social World, Essays and Experiments*, New York: McGraw-Hill.

182   REFERENCES

Mill J.S. (1909 [1848]), *Principles of Political Economy with some of their Applications to Social Philosophy*, 7th edition, London: Longmans, Green and Co.

Minar N., Burkhart R., Langton C., Askenazi M. (1996), The Swarm simulation system: a toolkit for building multi-agent simulations, Technical report 96-06-042, Santa Fe, NM: Santa Fe Institute, http://www.swarm.org (accessed March 24, 2015).

Nebel M. (2011), Implementation and analysis of "satisficing" as a model for farmers' decision- making in an agent-based model of groundwater over-exploitation, University of Osnabrück, Germany.

Neu W. (2008), Making economic sense of brain models: a survey and interpretation of the literature, *Journal of Bioeconomics*, 10(2): 165–192.

Newell B.R. (2005), Re-visions of rationality?, *Trends in Cognitive Sciences*, 9(1): 11–15.

Newman M., Barabási A.L., Watts D.J. (2006), *The Structure and Dynamics of Networks*, Princeton, NJ: Princeton University Press.

North M.J., Collier N.T., Ozik J., et al. (2013), Complex adaptive systems modeling with Repast Simphony, *Complex Adaptive Systems Modeling*, 1(3): 1–26.

Nowak A., Kùs M., Urbaniak J., Zarycki T. (2000), Simulating the coordination of economic decisions, *Physica A*, 287: 613–630.

Ostrom T. (1988), Computer simulation: the third symbol system, *Journal of Experimental Social Psychology*, 24: 381–392.

Padgham L., Scerri D., Jayatilleke G., Hickmott S. (2011), Integrating BDI reasoning into agent based modeling and simulation, in Jain S., Creasey R.R., Himmelspach J., White K.P., Fu M. (eds), *Proceedings of the 2011 Winter Simulation Conference*, Piscataway, NJ: IEEE, 345–356.

Parunak H.V., Bisson R., Brueckner S., Matthews R., Sauter J. (2006), Modeling agent emotions, in Sallach D.L., Macal C.M., North M.J. (eds), *Proceedings of the Agent 2006 Conference on Social Agents: Results and Prospects*, Chicago, IL: Argonne National Laboratory/The University of Chicago, 137–144.

Popper K. (1935 [2002]), *The Logic of Scientific Discovery*, Routledge.

Prigogine, I., and Stengers, I. (1979), *La nouvelle alliance: métamorphose de la science*, Paris: Gallimard.

Qin Y., Bothell D., Anderson J.R. (2007), ACT-R meets fMRI, in Zhong N., Liu J., Yao Y., Wu J., Lu S., Li K. (eds), *Web Intelligence Meets Brain Informatics*, Lecture Notes in Computer Science 4845, Berlin Heidelberg: Springer, 205–222.

Quinlan J.R. (1986), Induction of decision trees, *Machine Learning*, 1: 81–106.

Railsback S.F., Lytinen S.L., Jackson S.K. (2006), Agent-based simulation platforms: review and development recommendations, *Simulation*, 82(9): 609–623.

Rand W. (2006), Machine learning meets agent-based modeling: when not to go to a bar, in Sallach D.L., Macal C.M., North M.J. (eds), *Proceedings of the Agent 2006 Conference on Social Agents: Results and Prospects*, Chicago, IL: Argonne National Laboratory/The University of Chicago, 51–58.

Rapoport A. and Suleiman R. (1993), Incremental contribution in step-level public goods games with asymmetric players, *Organizational Behavior and Human Decision Processes*, 55: 171–194.

Richetin J., Sengupta A., Perugini M., et al. (2010), A micro-level simulation for the prediction of intention and behavior, *Cognitive Systems Research*, 11(2): 181–193.

Rieskamp J. and Hoffrage U. (1999), When do people use simple heuristics, and how can we tell? in Gigerenzer G., Todd P.M., the ABC Research Group (eds), *Simple Heuristics that Make Us Smart*, New York: Oxford University Press, 141–167.

Rieskamp J.R. and Otto P.E. (2006), SSL: a theory of how people learn to select strategies, *Journal of Experimental Psychology: General*, 135(2): 207.

Robinson D.T., Brown D.G., Parker D.C., et al. (2007), Comparison of empirical methods for building agent-based models in land use science, *Journal of Land Use Science*, 2(1): 31–55.

Ron S. (ed.) (2006), *Cognition and Multi-Agent Interaction: From Cognitive Modeling to Social Simulation*, Cambridge: Cambridge University Press.

Ross D. (2008a), Economics, cognitive science and social cognition, *Cognitive Systems Research*, 9(1–2): 125–135.

Ross D. (2008b), Two styles of neuroeconomics, *Economics and Philosophy*, 24: 373–383.

Roth A.E. and Erev I. (1995), Learning in extensive-form games: experimental data and simple dynamic models in the intermediate term, *Games and Economic Behavior*, Special Issue: Nobel Symposium, 8(1): 164–212.

Sabater J., Paolucci M., Conte R. (2006), Repage: REPutation and imAGE among limited autonomous partners, *Journal of Artificial Societies and Social Simulation*, 9(2).

Sawyer R.K. (2005), *Social Emergence—Societies as Complex Systems*, Cambridge: Cambridge University Press.

Schelling T. (1971), Dynamic models of segregation, *Journal of Mathematical Sociology*, 1: 143–186.

Schelling T. (1998), Social mechanisms and social dynamics, in Hedström P. and Swedberg R. (eds), *Social Mechanisms: An Analytical Approach to Social Theory*, Cambridge: Cambridge University Press, 32–44.

Schmitz C. (2012), *LimeSurvey: An Open Source Survey Tool*, Hamburg, Germany: LimeSurvey Project Hamburg.

Science and Technology Select Committee (2011), *Behaviour Change, Report of the Authority of the House of Lords*, London: The Stationery Office Limited.

Searle S.R., Casella G., McCulloch C.E. (1992), *Variance Components*, New York: John Wiley & Sons, Inc.

Selten R. (1998), Aspiration adaptation theory, *Journal of Mathematical Psychology*, 42: 191–214.

Shoam Y., Powers R., Grenager T. (2007), If multi-agent learning is the answer, what is the question? *Artificial Intelligence*, 171: 365–377.

Shubik M. (1996), Simulations, models and simplicity, *Complexity*, 2(1): 60–60.

Simon H.A. (1956), Rational choice and the structure of the environment, *Psychological Review*, 63(2): 129–138.

Smith V.L. (2003), Constructivist and ecological rationality in economics, *American Economic Review*, 93(3): 465–508.

Soll J.B. and Larrick R.P. (2009), Strategies for revising judgment: how (and how well) people use others' opinions, *Journal of Experimental Psychology: Learning, Memory, and Cognition*, 35(3): 780.

Squazzoni F. (2012), *Agent-Based Computational Sociology*, Chichester: John Wiley & Sons, Ltd.

StataCorp (2013), *Stata Statistical Software: Release 13*, College Station, TX: StataCorp LP.

Stinchcombe A.L. (1991), The conditions of fruitfulness of theorizing about mechanisms in social science, *Philosophy of the Social Sciences*, 21(3): 367–388.

Swamy P.A.V.B. (1970), Efficient inference in a random coefficient regression model, *Econometrica*, 38: 311–323.

Terna P. (2000), Economic experiments with Swarm: a neural network approach to the self-development of consistency in agents' behavior, in Luna F. and Stefansson B. (eds), *Economic Simulations in Swarm: Agent-Based Modelling and Object Oriented Programming*, Dordrecht/London: Kluwer.

Thorne B.C., Bailey A.M., Peirce S.M. (2007), Combining experiments with multicell agent-based modeling to study biological tissue patterning, *Briefings in Bioinformatics*, 8(4): 245–257.

Tobin J. (1958), Estimation of relationships for limited dependent variables, *Econometrica*, 26: 24–36.

Todd P.M. and Dieckmann A. (2005), Heuristics for ordering cue search in decision making, in Saul L.K., Weiss Y., Bottou L. (eds), *Advances in Neural Information Processing Systems*, 17, Cambridge, MA: MIT Press, 1393–1400.

Tubaro P. and Casilli A.A. (2010), "An ethnographic seduction": how qualitative research and agent-based models can benefit each other, *Bulletin de Méthodologie Sociologique*, 106(1): 59–74.

Van Den Berg A. (1998), Is sociological theory too grand for social mechanisms? in Hedström P. and Swedberg R. (eds), *Social Mechanisms: An Analytical Approach to Social Theory*, Cambridge: Cambridge University Press, 204–237.

Verbeek H. (2012), *A Guide to Modern Econometrics*, 4th edition, Chichester: John Wiley & Sons, Ltd.

Von Bastian C.C., Locher A., Ruflin M. (2013), Tatool: a Java-based open-source programming framework for psychological studies, *Behavior Research Methods*, 45(1): 108–115.

Vriend N.J. (2000), An illustration of the essential difference between individual and social learning, and its consequences for computational analysis, *Journal of Economic Dynamics & Control*, 24: 1–19.

Wasserman S. and Faust K. (1994), *Social Network Analysis: Methods and Applications*, Cambridge: Cambridge University Press.

Watts D.J. (1999), *Small Worlds: The Dynamics of Networks between Order and Randomness*, Princeton, NJ: Princeton University Press.

Watts D.J. and Strogatz S. (1998), Collective dynamics of "small-world" networks, *Nature*, 393: 440–442.

Weisbuch G., Kirman A., Herreiner D. (2000), Market organisation and trading relationships, *Economic Journal*, 110(463): 411–436.

West D.B. (2000), *Introduction to Graph Theory*, 2nd edition, Englewood Cliffs, NJ: Prentice-Hall.

Wilensky U. (1999), *NetLogo, Center for Connected Learning and Computer-Based Modeling*, Evanston, IL: Northwestern University.

Willemsen M.C. and Johnson E.J. (2010), Visiting the decision factory: observing cognition with MouselabWEB and other information acquisition methods, in Schulte-Mecklenbeck M., Kühberger A., Ranyard R. (eds), *A Handbook of Process Tracing Methods for Decision Making*, New York: Taylor & Francis, 21–42.

Wittmann T. (2008), *Agent-Based Models of Energy Investment Decisions*, Heidelberg: Physica-Verlag.

Zellner A. (1962), An efficient method of estimating seemingly unrelated regressions and tests for aggregation bias, *Journal of the American Statistical Association*, 57: 348–368.

Zoethout K. and Jager W. (2009), A conceptual linkage between cognitive architectures and social interaction, *Semiotica*, 175(1/4): 317–333.

Index